Distinctive Aspects of Molybdenum Disulfide

Edited by Dhanasekaran Vikraman and Hyun-Seok Kim

Published in London, United Kingdom

Distinctive Aspects of Molybdenum Disulfide
http://dx.doi.org/10.5772/intechopen.1005922
Edited by Dhanasekaran Vikraman and Hyun-Seok Kim

Contributors
Changjian Zhou, Daniel Olaya-Cortes, Dhanasekaran Vikraman, Hyun-Seok Kim, Jiaming Hu, Liqiao Chen, Lun Wang, Meilan Pan, Min Sup Choi, Muhammad Sabbtain Abbas, Rongfei Wei, Wanting Yang, Wensheng Xie, Xiangling Tian, Xiuzhen Wei, Yasir Hassan, Yenny Hernandez, Zichao Ma

First published in London, United Kingdom, 2025 by IntechOpen
IntechOpen is the global imprint of INTECHOPEN LIMITED, registered in England and Wales, registration number: 11086078, 167-169 Great Portland Street, London, W1W 5PF, United Kingdom

For EU product safety concerns: IN TECH d.o.o., Prolaz Marije Krucifikse Kozulić 3, 51000 Rijeka, Croatia, info@intechopen.com or visit our website at intechopen.com.

British Library Cataloguing-in-Publication Data
A catalogue record for this book is available from the British Library

Distinctive Aspects of Molybdenum Disulfide
Edited by Dhanasekaran Vikraman and Hyun-Seok Kim
p. cm.
Print ISBN 978-1-83634-881-8
Online ISBN 978-1-83634-880-1
eBook (PDF) ISBN 978-1-83634-882-5

Meet the editors

Prof. Dhanasekaran Vikraman is an Associate Professor in the Division of Electronics and Electrical Engineering at Dongguk University in Seoul, South Korea. He earned his Bachelor's degree from Mannai Rajagopalaswamy Government Arts College in Mannargudi, which is affiliated with Bharathidasan University in Tiruchirappalli, India. He subsequently completed both his Master's and Ph.D. degrees in the Department of Physics at Alagappa University, India. He has held various prestigious positions, including a visiting scientist role at KIST in South Korea, a Marie Curie Experienced Researcher fellowship in the Department of Physics at the Aristotle University of Thessaloniki in Greece, and postdoctoral positions at Sejong University and Ajou University in South Korea. His research interests primarily focus on the low-cost synthesis of inorganic low-dimensional materials for applications in electronics and energy devices.

Prof. Hyun-Seok Kim received his M.Sc. and Ph.D. degrees in Electrical Engineering from the University of Wisconsin–Madison, USA, in 2003 and 2007, respectively. Following this, he worked as a postdoctoral researcher at the same institution, where he focused on the design and fabrication of nanoelectromechanical single-electron transistors. From 2008 to 2011, he served as a Senior R&D Engineer at Intel Corporation in the USA. He is currently a Full Professor in the Division of Electronics and Electrical Engineering at Dongguk University-Seoul, Seoul, South Korea. His primary research interests include the synthesis of nanocomposites and their applications in sensors and energy-efficient devices.

Contents

Preface

Molybdenum disulfide (MoS_2) materials have garnered significant interest in recent years due to their remarkable versatility and potential applications across a wide range of fields. Their unique properties position them as effective solutions in various areas, including microbial disinfection, semiconductor technology, pollutant degradation, photovoltaics, water splitting, self-cleaning surfaces, cancer treatment, and photocatalysis, among others. One of the key features of MoS_2 is its tunable band gap, which allows it to be engineered for various electronic applications. This property, combined with its distinctive layered structure, makes MoS_2 a compelling alternative to other two-dimensional materials traditionally used in semiconductor devices. For instance, it shows promise in the development of optical sensors, biosensors, and electrochemical biosensors, where high sensitivity and responsive characteristics are essential.

Moreover, MoS_2 plays a crucial role in the energy sector, being utilized in batteries for enhanced performance, solar cells to improve efficiency, and microwave and terahertz applications due to its unique electronic and optical properties. On a nanoscale level, MoS_2 exhibits excellent potential in the realms of spintronics and magnetoresistance, which are pivotal in advancing next-generation electronic devices. MoS_2 has also garnered particular attention in the medical field, particularly for cancer diagnosis and treatment. Its advantages, including high photothermal conversion efficiency, an extensive surface area, and favorable biocompatibility, make MoS_2 a promising candidate for innovative cancer treatment strategies. Researchers are exploring its ability to deliver localized treatment while minimizing damage to surrounding healthy tissue. In addition, nano-sized MoS_2 particles, which can exhibit various shapes and sizes, have proven to be effective lubricant additives. Their unique characteristics enhance performance in lubrication applications, reducing friction and wear in mechanical systems.

This book delves deeply into the various properties of MoS_2, with each chapter focusing on a specific application, exploring the challenges faced, and presenting insights into the future trajectory of this versatile material in technological advancements.

Dhanasekaran Vikraman and Hyun-Seok Kim
Division of Electronics and Electrical Engineering,
Dongguk University-Seoul,
Seoul, Korea

Chapter 1

Introductory Chapter: Molybdenum Disulfide – Characteristics and Features

Dhanasekaran Vikraman and Hyun-Seok Kim

1. Introduction

Transition-metal dichalcogenides (TMDs), which are fascinating analogues to graphene, have garnered significant attention in the scientific community due to their remarkable potential for fundamental research and innovative technological applications [1, 2]. The unique coordination of transition metals by chalcogens, along with the varying stacking sequences of multiple layers of TMD materials, gives rise to a rich array of electronic properties. These anisotropic characteristics of TMDs have been effectively leveraged for advanced energy storage and conversion, as well as in the fields of electrochemical catalysis and sensing [3, 4]. Among the diverse family of TMDs, molybdenum disulfide (MoS_2) stands out as a heavily investigated material in recent years [5, 6]. Its distinct triple-layered configuration—consisting of a Mo-layer sandwiched between two S-layers—exhibits fascinating properties [7, 8]. The S-Mo-S monolayers are held together by relatively weak van der Waals forces, allowing for their separation and manipulation [9, 10]. Notably, MoS_2 features a direct band gap of ~1.8 eV in its monolayer form, and its layer-dependent band structure effectively resolves the gapless conductance issues associated with graphene, thereby underscoring its significance in both scientific inquiry and industrial applications [11–13].

2. Characteristics

As illustrated in **Figure 1a**, a single-layer MoS_2 typically exhibits either an octahedral or a trigonal prismatic coordination phase [14]. In this layered architecture, each molybdenum atom is coordinated to six sulfur atoms. The three most frequently faced polymorphs of MoS_2, designated as 1 T, 2H, and 3R, reflect various atomic arrangements as depicted in **Figure 1b**. The letter suggests the type of symmetry: R for rhombohedral (C_{53v} group), H for hexagonal (D_{3h} group), and T for tetragonal (D_{3d} group), while the numeral signifies the number of layers in the crystallographic unit cell [15]. Among these, the 2H-polytype is recognized as highly stable and often found in naturally occurring MoS_2 crystals, such as molybdenite. In contrast, the 1 T and 3R phases are considered metastable and predominantly observed in synthetic MoS_2 [16, 17]. In the layered structure, three sulfur and three molybdenum atoms are joined to form a hexatomic ring. In the

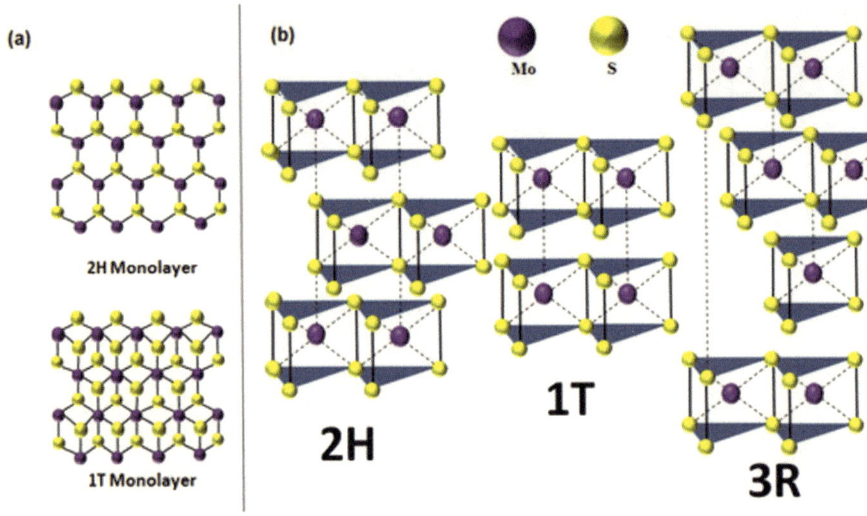

Figure 1.
(a) Atomic arrangement of 2H and 1 T MoS_2 monolayer. (b) 2H, 1 T and 3R phase MoS_2 crystal forms (reproduced and adapted from Ref. [14]).

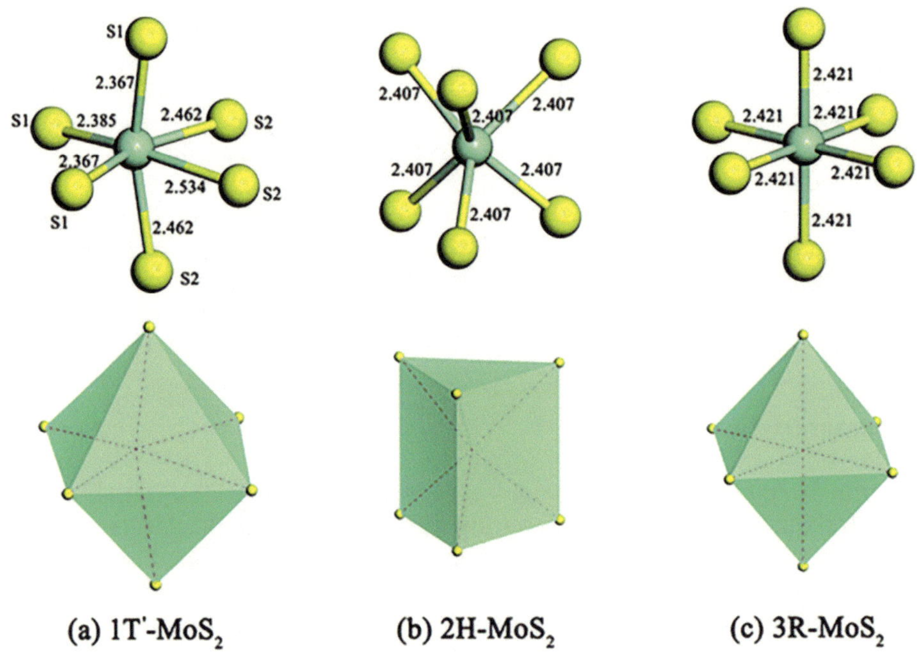

Figure 2.
The unit cell construction and polyhedron of (a) 1 T′-MoS_2; (b) 2H-MoS_2; (c) 3R-MoS_2 crystal phases (reproduced and adapted from Ref. [18]).

rhombohedral and hexagonal structures, the hexatomic ring forms a regular hexagon in nature, whereas the tetragonal structure creates a distorted hexagon [18]. MoS_2 exhibits octahedral coordination in the 3R and 1 T phases, while trigonal prismatic coordination is exhibited in the 2H-MoS_2 (**Figure 2**) [18]. 3R and 2H

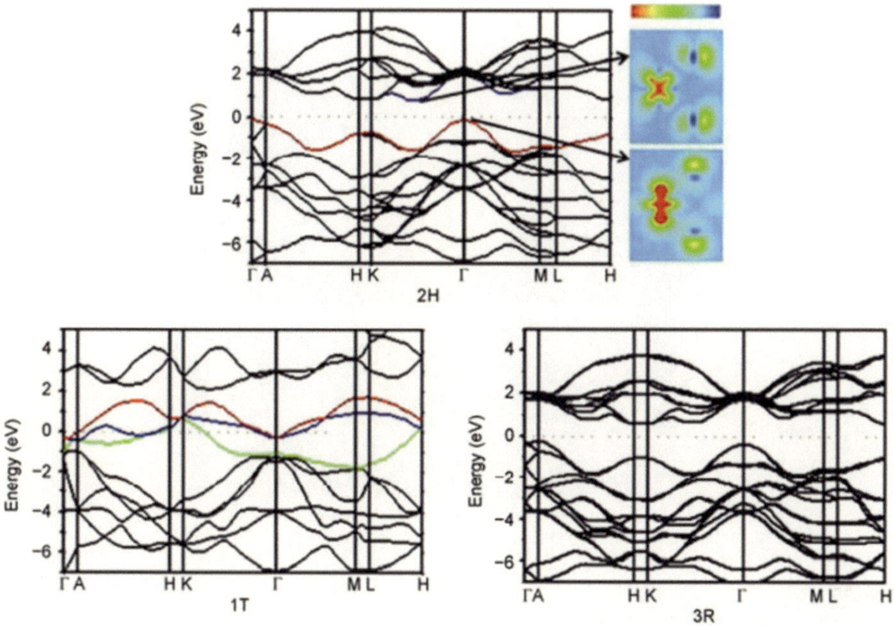

Figure 3.
Band structures of relaxed 2H-, 3R- and 1 T-MoS_2 (Fermi energy = 0 eV denoted by dotted line) (reproduced and adapted from Ref. [23]).

phases MoS_2 show identical Mo–S bond lengths, whereas 1 T′-MoS_2 exhibits different Mo–S bond lengths [19, 20]. In addition, **Figure 2a-c** describes the central variation between the three phases of MoS_2 polyhedron structures [18]. Notably, metastable MoS_2 can transition to the more stable 2H-polytype simply through post-heat treatments [15, 21]. The lattice parameters for the 2H, 3R, and 1 T forms were recorded from the theoretical results as follows: a = 5.81 Å, b = 3.24 Å and c = 5.40 Å for 1 T; a = b = 3.19 Å and c = 18.28 Å for 3R; a = b = 3.18 Å and c = 12.38 Å for 2H [18, 22]. **Figure 3** describes the band structures of relaxed 2H, 3R, and 1 T phases [23]. Authors have explored the valence band maximum and the conduction band minimum position for the 1 T, 2H and 3R phases.

3. Features

The unique characteristics of MoS_2 as a naturally occurring mineral, combined with its superior properties, render it highly suitable for a multitude of applications across various industrial sectors [24, 25]. Recent theoretical studies have illuminated the pivotal role of sulfur-terminated edges of MoS_2 in acting as active sites for the various electrochemical and photo-catalytic activities [26, 27]. Consequently, several strategies were proposed to enhance the electrochemical properties of various TMDs, including MoS_2, which involves the fabrication of materials with an increased surface area of edge-exposed sites compared to the basal planes. **Figure 4** illustrates the energy landscape between the 2H and 1 T crystals as a function of the charge doping and reaction coordinate [28]. In this context, the reaction coordinate represents the Euclidean gap along the phase shift path that connects the 2H and 1 T arrangements [29].

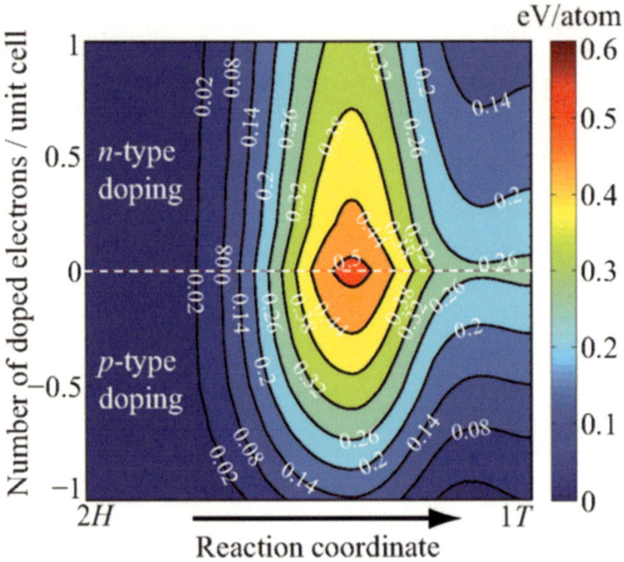

Figure 4.
Energy landscape as a function of charge doping and reaction coordinate along the path between the 2H and 1 T crystal forms (reproduced and adapted from Ref. [28]).

Furthermore, it is well established that the properties of MoS_2 materials in their bulk form diverge significantly from those of their exfoliated counterparts. Another critical determinant of device properties is the material's electrical conductivity, which facilitates better communication between the electron source and the active edge sites, thereby enhancing device efficiency. Here's a breakdown of its key characteristics and features:

- Each layer consists of Mo atoms bonded to six S atoms in a trigonal prismatic arrangement.
- The thickness of 2D MoS_2 monolayer is ~0.65 nm, and it is held by weak van der Waals forces to make bulk MoS_2.
- Indirect bandgap semiconducting nature is exhibited in the bulk- and multi-layer MoS_2, and it can be transformed into direct bandgap by changing to mono-layer MoS_2.
- Layers of MoS_2 are held together by weak van der Waals forces, making it easy to exfoliate into thin sheets (nanosheets) similar to graphene.
- MoS_2 is a relatively soft material with a hardness of about 1–2 on the Mohs scale, which contributes to its excellent lubricating properties.
- MoS_2 is chemically stable at high temperatures and does not easily oxidize. It is also resistant to corrosion in many environments.
- MoS_2 has excellent self-lubricating properties owing to its layer architecture, which reduces friction and wear when used as a lubricant.

- MoS_2 helps reduce wear resistance in mechanical systems, enhancing the longevity of moving parts in engines, aerospace, and other machinery.

- MoS_2 is widely used as a solid lubricant due to its low friction coefficient, particularly in high-temperature and high-pressure environments.

- MoS_2 has been used in batteries and supercapacitors, where it can serve as an electrode material to enhance performance.

- MoS_2 has been employed as a catalyst in petroleum refining and the water splitting kinetics.

- MoS_2 is being investigated as a material for thin-film solar cells due to its excellent electronic properties. It can potentially contribute to more efficient and cost-effective solar energy production.

- MoS_2 can be used in coatings to protect metals from corrosion, especially in harsh environments like marine and industrial applications, where it acts as a barrier, preventing wear and corrosion.

In summary, molybdenum disulfide's unique physical and chemical properties make it versatile in several advanced industrial, technological, and research applications.

Author details

Dhanasekaran Vikraman* and Hyun-Seok Kim
Division of Electronics and Electrical Engineering, Dongguk University-Seoul, Seoul, Republic of Korea

*Address all correspondence to: v.j.dhanasekaran@gmail.com

References

[1] Wang QH, Kalantar-Zadeh K, Kis A, Coleman JN, Strano MS. Electronics and optoelectronics of two-dimensional transition metal dichalcogenides. Nature Nanotechnology. 2012;7:699

[2] Li H, Tsai C, Koh AL, Cai L, Contryman AW, Fragapane AH, et al. Activating and optimizing MoS_2 basal planes for hydrogen evolution through the formation of strained Sulphur vacancies. Nature Materials. 2016;**15**(1):48-53

[3] Vikraman D, Akbar K, Hussain S, Yoo G, Jang JY, Chun SH, et al. Direct synthesis of thickness-tunable MoS_2 quantum dot thin layers: Optical, structural and electrical properties and their application to hydrogen evolution. Nano Energy. 2017;**35**:101-114

[4] Brivio J, Alexander DT, Kis A. Ripples and layers in ultrathin MoS_2 membranes. Nano Letters. 2011;**11**(12):5148-5153

[5] Vikraman D, Karuppasamy K, Hussain S, Kathalingam A, Sanmugam A, Jung J, et al. One-pot facile methodology to synthesize MoS_2-graphene hybrid nanocomposites for supercapacitors with improved electrochemical capacitance. Composites Part B: Engineering. 2019;**161**:555-563

[6] Hussain S, Shaikh SF, Vikraman D, Mane RS, Joo O-S, Naushad M, et al. High-performance platinum-free dye-sensitized solar cells with molybdenum Disulfide films as counter electrodes. ChemPhysChem. 2015;**16**(18):3959-3965

[7] Ye G, Gong Y, Lin J, Li B, He Y, Pantelides ST, et al. Defects engineered monolayer MoS_2 for improved hydrogen evolution reaction. Nano Letters. 2016;**16**(2):1097-1103

[8] Singha SS, Rudra S, Mondal S, Pradhan M, Nayak AK, Satpati B, et al. Mn incorporated MoS_2 nanoflowers: A high performance electrode material for symmetric supercapacitor. Electrochimica Acta. 2020;**338**:135815

[9] Geim AK, Grigorieva IV. Van der Waals heterostructures. Nature. 2013;**499**:419

[10] Hussain S, Akbar K, Vikraman D, Karuppasamy K, Kim H-S, Chun S-H, et al. Synthesis of $MoS_{2(1-x)}Se_{2x}$ and $WS_{2(1-x)}Se_{2x}$ alloys for enhanced hydrogen evolution reaction performance. Inorganic Chemistry Frontiers. 2017;**4**(12):2068-2074

[11] Vikraman D, Hussain S, Truong L, Karuppasamy K, Kim HJ, Maiyalagan T, et al. Fabrication of MoS_2/WSe_2 heterostructures as electrocatalyst for enhanced hydrogen evolution reaction. Applied Surface Science. 2019;**480**:611-620

[12] Ganatra R, Zhang Q. Few-layer MoS_2: A promising layered semiconductor. ACS Nano. 2014;**8**(5):4074-4099

[13] Vikraman D, Arbab AA, Hussain S, Shrestha NK, Jeong SH, Jung J, et al. Design of WSe_2/MoS_2 heterostructures as the counter electrode to replace Pt for dye-sensitized solar cell. ACS Sustainable Chemistry & Engineering. 2019;7(15):13195-13205

[14] Mouloua D, Kotbi A, Deokar G, Kaja K, El Marssi M, El Khakani MA, et al. Recent Progress in the synthesis of MoS_2 thin films for sensing, photovoltaic and Plasmonic applications: A review. Materials. 2021;**14**(12):3283

[15] Zhao W, Pan J, Fang Y, Che X, Wang D, Bu K, et al. Metastable

MoS_2: Crystal structure, electronic band structure, synthetic approach and intriguing physical properties. Chemistry – A European Journal. 2018;**24**(60):15942-15954

[16] Wypych F, Schöllhorn R. 1T-MoS_2, a new metallic modification of molybdenum disulfide. Journal of the Chemical Society, Chemical Communications. 1992;**19**:1386-1388

[17] Vikraman D, Hussain S, Ali M, Karuppasamy K, Santhoshkumar P, Hwang J-H, et al. Theoretical evaluation and experimental investigation of layered 2H/1T-phase MoS_2 and its reduced graphene-oxide hybrids for hydrogen evolution reactions. Journal of Alloys and Compounds. 2021;**868**:159272

[18] Zhao Z-Y, Liu Q-L. Study of the layer-dependent properties of MoS_2 nanosheets with different crystal structures by DFT calculations. Catalysis Science and Technology. 2018;**8**(7):1867-1879

[19] Fan X-L, Yang Y, Xiao P, Lau W-M. Site-specific catalytic activity in exfoliated MoS_2 single-layer polytypes for hydrogen evolution: Basal plane and edges. Journal of Materials Chemistry A. 2014;**2**(48):20545-20551

[20] Hussain S, Vikraman D, Truong L, Akbar K, Rabani I, Kim HS, et al. Facile and cost-effective growth of MoS_2 on 3D porous graphene-coated Ni foam for robust and stable hydrogen evolution reaction. Journal of Alloys and Compounds. 2019;**788**:267-276

[21] Vikraman D, Hussain S, Karuppasamy K, Santhoshkumar P, Kathalingam A, Jung J, et al. Fabrication of asymmetric supercapacitors using molybdenum dichalcogenide nanoarray structures. International Journal of Energy Research. 2022;**46**(13):18410-18425

[22] Schonfeld B, Huang JJ, Moss SC. Anisotropic mean-square displacements (MSD) in single-crystals of 2H- and 3R-MoS_2. Acta Crystallographica Section B. 1983;**39**(4):404-407

[23] Chen X, Chen Z, Li J. Critical electronic structures controlling phase transitions induced by lithium ion intercalation in molybdenum disulphide. Chinese Science Bulletin. 2013;**58**(14):1632-1641

[24] Sun T, Tang Z, Zang W, Li Z, Li J, Li Z, et al. Ferromagnetic single-atom spin catalyst for boosting water splitting. Nature Nanotechnology. 2023;**18**(7):763-771

[25] Faizan M, Hussain S, Islam M, Kim J-Y, Han D, Bae J-H, et al. MoO3@MoS2 Core-Shell Structured Hybrid Anode Materials for Lithium-Ion Batteries. Nanomaterials. 2022;**12**(12):2008

[26] Hezam A, Alkanad K, Bajiri MA, Strunk J, Takahashi K, Drmosh QA, et al. 2D/1D MoS_2/TiO_2 heterostructure photocatalyst with a switchable CO_2 reduction product. Small Methods. 2023;**7**(1):2201103

[27] Thomas N, Mathew S, Nair KM, O'Dowd K, Forouzandeh P, Goswami A, et al. 2D MoS_2: Structure, mechanisms, and photocatalytic applications. Materials Today Sustainability. 2021;**13**:100073

[28] Zhuang HL, Johannes MD, Singh AK, Hennig RG. Doping-controlled phase transitions in single-layer MoS_2. Physical Review B. 2017;**96**(16):165305

[29] Gan X, Lee LYS, Wong KY, Lo TW, Ho KH, Lei DY, et al. 2H/1T phase transition of multilayer MoS_2 by electrochemical incorporation of S vacancies. ACS Applied Energy Materials. 2018;**1**(9):4754-4765

Chapter 2

Piezochemistry of Molybdenum Disulfide

Xiuzhen Wei, Jiaming Hu, Liqiao Chen and Meilan Pan

Abstract

Molybdenum disulfide (MoS_2), a representative two-dimensional (2D) transition metal dichalcogenide, has attracted growing attention due to its intrinsic piezoelectric properties in monolayer and few-layer configurations. This chapter provides a comprehensive overview of the fundamental mechanisms, synthesis strategies, and structural engineering techniques used to enhance the piezoelectric performance of MoS_2. Key topics include the origin of piezoelectricity in non-centrosymmetric MoS_2, the role of doping, strain modulation, and interface engineering, as well as the development of nanostructures such as nanoribbons, and nanosheets. Fabrication methods ranging from mechanical exfoliation to chemical vapor deposition and atomic layer deposition are discussed in the context of material quality and device applicability. The practical implications of MoS_2 piezoelectric nanomaterials are explored across a wide range of applications, including energy harvesting, flexible electronics, biomedical sensing, and piezocatalysis. Finally, current challenges and future directions are highlighted to guide further research in this rapidly evolving field. This chapter aims to provide a timely and critical understanding of MoS_2-based piezoelectric systems, facilitating their advancement toward high-performance, scalable, and multifunctional applications in next-generation technologies.

Keywords: molybdenum disulfide, piezoelectric nanomaterials, 2D materials, energy harvesting, piezocatalysis

1. Introduction

The emergence of two-dimensional (2D) materials has opened new frontiers in the development of flexible, efficient, and multifunctional devices. Among these materials, molybdenum disulfide (MoS_2) has attracted significant attention due to its tunable electronic structure, strong spin–orbit coupling, and exceptional mechanical flexibility. Beyond its semiconducting and catalytic properties, MoS_2 exhibits remarkable piezoelectric behavior when thinned down to monolayer or other non-centrosymmetric structures, positioning it as a promising candidate for piezotronic and piezocatalytic applications. With the latest research progress, the piezoelectric catalytic performance of oxygen-doped MoS_2 was explored, providing a new strategy for regulating the piezoelectricity of bulk materials that are non-piezoelectric transition needle disulfides and opening up a new horizon for the development of efficient piezoelectric catalysts [1].

In recent years, growing interest has focused on the piezoelectric properties of MoS_2 and their exploitation in fields such as energy harvesting, flexible electronics, biosensing, and environmental remediation. While early studies primarily addressed its electronic and photonic applications, the realization of strain-induced polarization and its coupling with catalytic and optoelectronic effects has given rise to the emerging field of piezochemistry. Including the piezoelectric-induced metal support interaction and the photo-piezoelectric synergy that achieves a hydrogen evolution efficiency of 15.6% [2, 3]. However, challenges remain in optimizing the piezoelectric response of MoS_2, particularly in terms of material fabrication, structural engineering, and interface design. Additionally, the mechanisms by which piezoelectricity modulate surface reactivity and charge transport require further elucidation to fully exploit its potential in piezocatalysis and multifunctional device platforms. Additionally, the mechanisms by which piezoelectricity modulates surface reactivity and charge transport require further elucidation through advanced techniques like 4D-STEM for atomic-scale strain mapping and femtosecond transient absorption spectroscopy for probing exciton dynamics [4, 5].

This chapter provides a comprehensive overview of the origin, mechanisms, fabrication methods, and enhancement strategies of piezoelectric MoS_2. We also highlight recent progress in nanostructural modifications, elemental doping, and heterostructure engineering, which have collectively advanced the development of MoS_2-based piezoelectric nanodevices. Furthermore, the chapter outlines the diverse applications of MoS_2 in energy conversion, biomedical sensing, nanoelectronics, optoelectronics, and environmental catalysis, offering insights into future directions for this rapidly evolving field.

2. Piezoelectricity in MoS_2: Background, piezoelectric coefficient, and mechanism

2.1 Background for the origin of piezoelectricity in MoS_2

Piezoelectricity is a fundamental property of non-centrosymmetric materials, where mechanical deformation induces electric polarization due to the displacement of charge centers. In conventional bulk materials, such as ceramics or certain crystals, this effect arises from the lack of inversion symmetry within their unit cell. In the case of transition metal dichalcogenides (TMDs) and particularly molybdenum disulfide (MoS_2), the piezoelectric response exhibits distinctive characteristics due to their unique atomic arrangements and layer-dependent symmetry.

Bulk MoS_2 crystallizes in a 2H phase with a centrosymmetric structure, where the inversion symmetry suppresses the piezoelectric response. However, when exfoliated into a monolayer, MoS_2 adopts a non-centrosymmetric D_3h point group, thereby enabling piezoelectric behavior (**Figure 1a**) [6]. This layer-dependent symmetry breaking introduces an intrinsic piezoelectricity that only appears in odd-layered MoS_2 (monolayer, trilayer, etc.), while even-layered MoS_2 retains inversion symmetry and remains non-piezoelectric [9].

This fundamental discovery positions monolayer MoS_2 as a promising candidate for next-generation nanoscale piezoelectric applications, including flexible energy harvesters and self-powered sensors, which benefit from its atomic thickness, mechanical flexibility, and semiconducting nature [10].

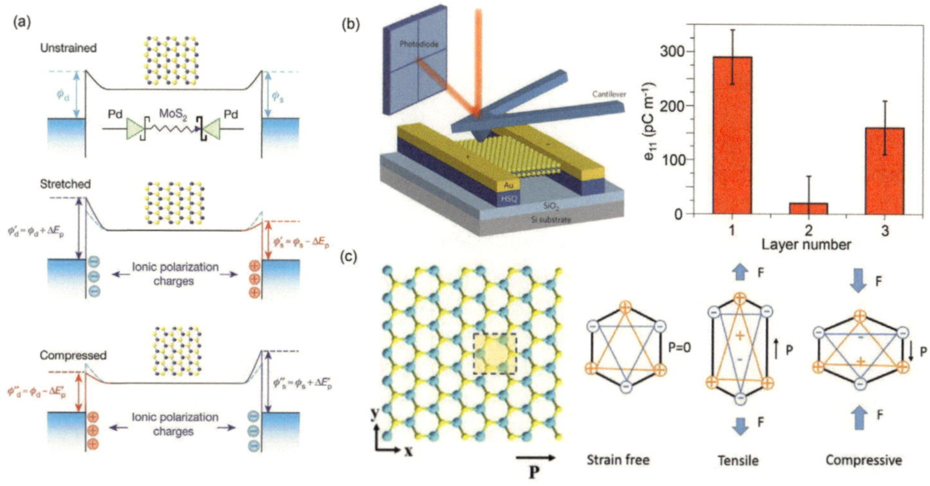

Figure 1.
Piezoelectrical properties of MoS_2. (a) The symmetrical modulation of carrier transport by strains under opposite drain bias in a bilayer device shows characteristics of a piezoresistive effect [6]. (b) To measure the in-plane piezoelectric stress, the MoS_2 film was suspended on two HSQ posts, clamped underneath by two Au electrodes, and the corresponding measured piezoelectric coefficient in one-, two-, and three-layer MoS_2 membranes, showing that only odd numbers of layers exhibit a significant coupling strength due to their broken inversion symmetry [7]. (c) Single-layer vertical view geometry of MoS_2 and the atomic structure model of piezoelectric potential generation in monolayer MoS_2 corresponding to the shadow area [8].

2.2 Piezoelectric coefficient (d_{11}) of MoS_2

The piezoelectric response of a material is commonly quantified by its piezoelectric coefficient (d_{11}), which represents the induced polarization per unit stress along a specific crystallographic direction. For monolayer MoS_2, both theoretical predictions and experimental measurements have demonstrated a non-negligible d_{11} value, which is significant given its two-dimensional nature.

First-principles calculations have predicted that monolayer MoS_2 exhibits a piezoelectric coefficient (d_{11}) in the range of 2.9 to 3.6 pm/V [11]. Experimental verification by Zhu et al. confirmed that the d_{11} value of monolayer MoS_2 is approximately 2.9×10^{-10} C m^{-1}, comparable to certain bulk piezoelectric materials (**Figure 1b**) [7]. Notably, the magnitude of d_{11} in MoS_2 is highly sensitive to factors such as layer number, strain conditions, and crystallographic orientation, with the piezoelectric effect vanishing in even-layered MoS_2 due to the restoration of inversion symmetry [12].

2.3 Mechanism of piezoelectricity in MoS_2

The piezoelectric behavior of monolayer MoS_2 originates from its unique trigonal prismatic structure, where molybdenum atoms are sandwiched between two sulfur atomic layers. When an in-plane mechanical strain is applied, there is an asymmetric displacement of the positively charged molybdenum and negatively charged sulfur atoms within the lattice (**Figure 1c**) [8]. This atomic distortion leads to a net dipole moment along the armchair direction of the crystal, generating an electric polarization proportional to the applied strain [13].

Furthermore, the piezoelectric effect in monolayer MoS_2 is characterized by its in-plane anisotropy. The polarization response is significantly stronger along the armchair direction, while it is negligible or absent along the zigzag direction [14]. This anisotropic behavior is governed by crystallographic orientation and atomic arrangement within the monolayer lattice.

3. Fabrication of piezoelectrical MoS_2

The fabrication of piezoelectric molybdenum disulfide (MoS_2) is a critical prerequisite for the development of advanced piezotronic and nanoelectromechanical systems (NEMS). Given that the piezoelectricity of MoS_2 arises from its non-centrosymmetric crystal structures—primarily in monolayer and certain few-layer configurations—precise control over material synthesis, crystallinity, thickness, and defect concentration is essential to achieve reliable and enhanced piezoelectric performance. Various synthesis and fabrication strategies have been developed to produce high-quality MoS_2 nanomaterials with tailored piezoelectric properties. These approaches can be broadly categorized into top-down exfoliation methods and bottom-up growth techniques [15–17].

3.1 Top-down exfoliation techniques

3.1.1 Mechanical exfoliation

Mechanical exfoliation, often referred to as the "Scotch tape method," is one of the earliest and simplest approaches used to obtain monolayer and few-layer MoS_2 from its bulk crystals. This technique relies on the weak van der Waals interactions between the layers of MoS_2, enabling facile separation under mechanical force (**Figure 2a**) [20]. While this method yields high-crystalline-quality flakes with excellent piezoelectric properties, its scalability is limited, restricting its applicability to fundamental research and prototyping.

3.1.2 Liquid phase exfoliation

Liquid phase exfoliation (LPE) provides a scalable route for the preparation of few-layer MoS_2 nanosheets by dispersing bulk MoS_2 crystals in suitable solvents under ultrasonic treatment (**Figure 2b**) [18]. The process often involves the use of surfactants or organic solvents to stabilize exfoliated layers. Although LPE facilitates mass production, the resulting MoS_2 flakes typically exhibit random size distributions, structural defects, and reduced piezoelectricity compared to mechanically exfoliated counterparts.

3.2 Bottom-up growth techniques

3.2.1 Chemical vapor deposition (CVD)

Chemical vapor deposition (CVD) has emerged as a highly promising technique for the synthesis of wafer-scale, monolayer MoS_2 with superior structural uniformity and tunable thickness (**Figure 2c**) [21, 22]. In a typical CVD process, Mo- or MoS_2-based precursors react with sulfur vapor at elevated temperatures on substrates such as SiO_2/Si, sapphire, or flexible polymers. CVD-grown monolayer MoS_2 exhibits

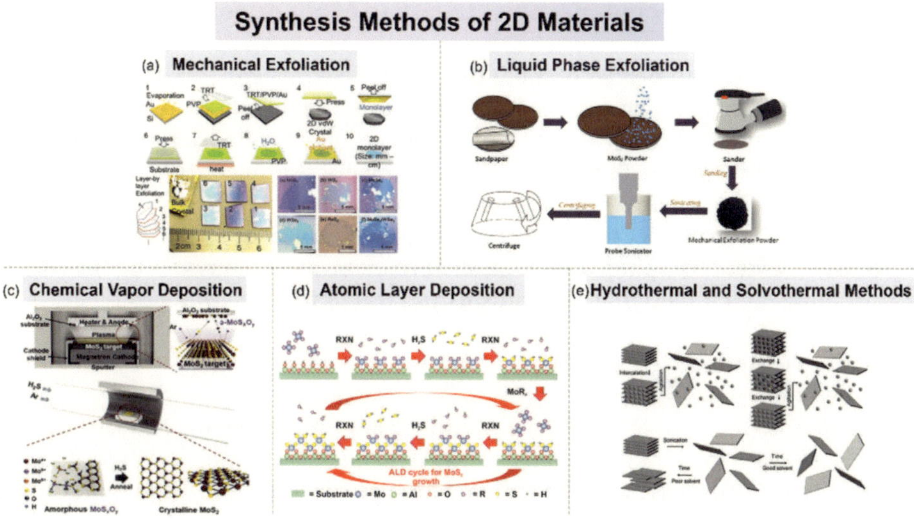

Figure 2.
Fabrication of piezoelectrical MoS_2. (a) Schematic representation of the mechanical exfoliation process, illustrating the step-by-step procedure to isolate monolayer or few-layer 2D materials from bulk crystals. (b) The typical green exfoliation process of single and few-layered MoS_2 nanosheets by a facile sanding and sonicating assisted method [18]. (c) Depiction of the PVD method, highlighting the use of a vapor-solid process for the synthesis of 2D materials with controlled thickness and uniformity. (d) Proposed scenario for MoS_2 film growth by ALD using $Mo(NMe_2)_4$ and H_2S in an A–B cycle; RXN: reaction [19]. (e) Illustration of solution-based methods, including exfoliation and intercalation, used for scalable production and functionalization of 2D nanosheets with precise layer control.

well-defined lattice structures, controlled grain boundaries, and minimal defects, enabling consistent piezoelectric properties over large areas. Furthermore, CVD allows precise engineering of domain orientation, which is essential for aligning piezoelectric dipoles in device architectures.

3.2.2 Atomic layer deposition (ALD)

Atomic layer deposition offers another promising approach for synthesizing ultra-thin MoS_2 films with atomic-scale thickness control (**Figure 2d**) [19]. ALD involves sequential exposure of substrates to molybdenum- and sulfur-containing precursors, facilitating layer-by-layer growth. Although still in developmental stages for MoS_2, ALD enables conformal coating over complex nanostructures, making it attractive for fabricating piezoelectric devices with non-planar geometries.

3.2.3 Hydrothermal and solvothermal methods

Hydrothermal synthesis provides a solution-based approach to grow MoS_2 nanostructures such as nanosheets, nanoribbons, and nanotubes. By controlling parameters such as temperature, reaction time, and precursor concentration, it is possible to tailor the morphology and thickness of MoS_2 (**Figure 2e**) [23]. These methods are particularly advantageous for producing curved or rolled structures, which can further enhance piezoelectric responses due to strain-induced polarization effects.

4. Key factors influencing piezoelectricity in MoS_2 nanomaterials

The piezoelectric properties of MoS_2 can be significantly enhanced and modulated through various engineering approaches, including strain modulation, elemental doping, structural modifications, and interface engineering (**Figure 3**). These methods enable precise control over the electronic and mechanical properties of MoS_2, making it highly suitable for applications in nanoelectromechanical systems (NEMS), energy harvesting, and flexible electronics. Below, we provide an in-depth analysis of key strategies—elemental doping, structural modifications, and interface engineering—by examining their underlying mechanisms and recent research findings.

4.1 Elemental doping

Elemental doping has emerged as a powerful strategy for tuning the piezoelectric properties of MoS_2 by altering its charge carrier dynamics, lattice structure, and polarization behavior [30]. Doping can be achieved through substitutional incorporation of transition metals, non-metal heteroatoms, or the introduction of vacancy defects [31].

4.1.1 Transition metal doping

Doping MoS_2 with transition metals such as iron(Fe) and cobalt (Co) has gained significant attention as an effective strategy to modulate its electronic structure, piezoelectric properties, and mechanical stability [32]. Transition metal doping alters the intrinsic properties of MoS_2 by introducing additional charge carriers, modifying the lattice strain, and enhancing polarization effects [33]. This approach has been widely explored in both theoretical and experimental studies to improve

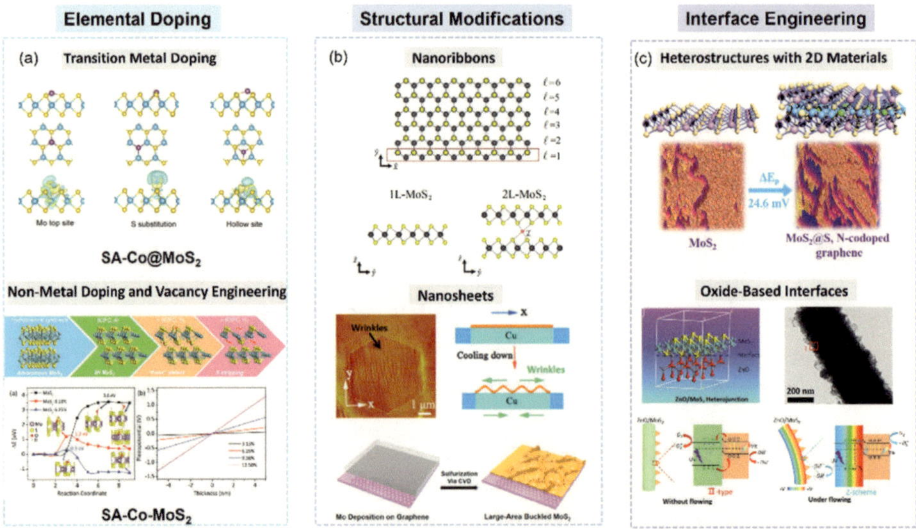

Figure 3.
Key factors influencing piezoelectricity in MoS_2 nanomaterials. (a) Illustration of SA-Co@MoS_2 [24] and S-vacancy MoS_2 [25]. (b) Schematic representation of MoS_2 nanoribbons [26] and nanosheets [27]. (c) Schematic representation of MoS_2@SNG [28] and ZnO@MoS_2 [29].

the piezoelectric efficiency of MoS_2-based NEMS and energy-harvesting devices [34]. DFT calculations have revealed that transition metal doping significantly impacts the band structure and charge distribution of MoS_2 [35]. For instance, Co-doped MoS_2 exhibits a shift in its density of states (DOS) near the Fermi level, leading to enhanced electron mobility and polarization [24]. Additionally, the introduction of transition metal dopants modifies the local bonding environment of MoS_2 [36]. Compared to pristine MoS_2, Fe and Co doping create localized lattice distortions, breaking inversion symmetry and thereby enhancing the in-plane piezoelectric coefficient (d11) [37].

Beyond electronic effects, transition metal doping introduces local strain variations that further boost piezoelectric performance [38]. The ionic radii of Fe and Co differ from that of Mo, causing localized compressive or tensile strain within the MoS_2 lattice. This strain results in increased atomic displacement under an external electric field, thereby amplifying the piezoelectric response. First-principles calculations predict that doping-induced strain can enhance the effective piezoelectric coefficient (d33) by up to 50% [11].

4.1.2 Non-metal doping and vacancy engineering

The introduction of oxygen or lattice, creating sulfur vacancies, or substitution with heteroatoms like nitrogen or phosphorus, can also impact the piezoelectric properties of MoS_2. Oxygen incorporation, for example, has been found to enhance the piezoelectric response by modifying the electron density distribution and stabilizing strain-induced polarization [39]. Similarly, controlled sulfur vacancies, induced through plasma treatment or annealing, have been shown to create localized dipoles that contribute to increased piezoelectric activity [25].

For example, the introduction of sulfur vacancies into MoS_2 is another promising approach to engineering piezoelectricity. Vacancies, particularly sulfur vacancies, create localized regions with altered bonding characteristics and electron distribution, which can significantly impact the material's piezoelectric properties. Mondal et al. reported that sulfur vacancies in MoS_2, created through plasma treatment or annealing processes, resulted in the formation of localized dipoles that enhanced the piezoelectric response of the material [25]. Their experiments demonstrated that sulfur vacancy engineering improved the effective piezoelectric coefficient, particularly under tensile strain, by promoting the redistribution of charge carriers and breaking the symmetry of the lattice. Additionally, sulfur vacancies were found to enhance the material's response to external stress, allowing for greater piezoelectric activity when subjected to mechanical deformation. This is because sulfur vacancies increase the atomic displacement in the MoS_2 lattice, thus amplifying the piezoelectric effect. This vacancy engineering approach offers a way to control and enhance the piezoelectric properties of MoS_2 without introducing complex dopants, making it a simpler yet effective strategy for optimizing the material for NEMS and energy-harvesting applications.

4.2 Structural modifications

Structural engineering techniques, including the fabrication of MoS_2 nanoribbons, nanoscrolls, and nanotubes, offer novel pathways to enhance piezoelectric properties by exploiting edge effects, curvature-induced strain, and quantum confinement phenomena.

4.2.1 MoS_2 nanoribbons

MoS_2 nanoribbons have recently attracted significant attention as an advanced form of MoS_2 with substantially enhanced piezoelectric properties compared to their monolayer counterparts. The unique morphology and edge characteristics of MoS_2 nanoribbons create distinct advantages for tuning the piezoelectric response, offering a new avenue for designing highly efficient piezoelectric devices at the nanoscale. Theoretical and experimental studies have shown that MoS_2 nanoribbons exhibit a significantly enhanced piezoelectric coefficient, primarily due to their asymmetric edge structure, which introduces additional polarization effects do not present in monolayers.

DFT calculations have revealed that the piezoelectric properties of MoS_2 nanoribbons are largely influenced by the asymmetric edge states, which lead to substantial strain redistribution within the material [40]. The broken symmetry at the edges of the nanoribbons results in localized charge accumulation, leading to an enhanced dipole moment under mechanical deformation. Specifically, DFT simulations predict that the in-plane piezoelectric coefficient of MoS_2 nanoribbons can be increased by nearly an order of magnitude compared to that of monolayer MoS_2, due to the enhanced polarization at the edges and strain effects caused by the nanostructuring. This strain redistribution is particularly pronounced in narrow ribbons, where the reduced dimensionality and edge asymmetry create more favorable conditions for enhanced piezoelectricity. Furthermore, these computational models also highlight the tunability of the piezoelectric response by controlling the width and edge configuration of the nanoribbons. For instance, narrow MoS_2 nanoribbons exhibit stronger edge polarization effects, whereas wider ribbons may display more uniform strain distribution, leading to a less pronounced piezoelectric enhancement (**Figure 3b**) [26]. This suggests that the geometric properties of the nanoribbon specifically their width and edge morphology are critical factors in optimizing the piezoelectric response.

4.2.2 MoS_2 nanosheet

MoS_2 nanosheets, as one of the most fundamental morphologies of MoS_2, exhibit layer-dependent piezoelectric properties that are critical for their piezocatalytic performance. When exfoliated to monolayer or few-layer structures, MoS_2 breaks inversion symmetry, enabling an intrinsic piezoelectric response to absent in bulk forms. The flat, planar morphology of nanosheets facilitates efficient strain transfer and polarization under mechanical deformation, which enhances charge separation and boosts surface catalytic activity. Density functional theory (DFT) studies and experimental observations confirm that monolayer MoS_2 nanosheets generate measurable piezoelectric polarization, with their performance significantly influenced by thickness, lateral size, and crystallinity [11].

Importantly, the presence of edge sites and surface defects—such as sulfur vacancies—further amplifies the piezoelectric response by introducing localized asymmetry and additional polarization centers. These structural features enhance dipole formation and promote more efficient electron–hole separation during catalytic processes [41]. Compared to bulk MoS_2, nanosheets possess a higher specific surface area and greater accessibility of active sites, both of which contribute to superior piezocatalytic activity. However, unlike nanoribbons or curved nanostructures, nanosheets lack pronounced edge asymmetry or curvature-induced strain, which somewhat limits their maximum achievable piezoelectric enhancement. Nonetheless, by tuning thickness, defect density, and strain orientation, the piezocatalytic performance of MoS_2 nanosheets can be significantly optimized for applications in environmental remediation and energy conversion.

4.3 Interface engineering

Interface engineering, particularly through the integration of MoS_2 with other 2D materials or transition metal oxides, has emerged as a crucial approach for optimizing piezoelectric properties. By forming van der Waals heterostructures, researchers have been able to manipulate charge transfer dynamics, stabilize piezoelectric performance under mechanical cycling, and reduce charge screening effects.

4.3.1 Heterostructures with 2D materials

The integration of MoS_2 with other 2D materials, such as hexagonal boron nitride (h-BN), graphene, and transition metal dichalcogenides (TMDs), has emerged as a highly effective strategy for enhancing and stabilizing piezoelectricity in nanoscale devices. By leveraging the unique electronic, mechanical, and interfacial properties of these heterostructures, researchers have demonstrated significant improvements in piezoelectric performance, charge transfer efficiency, and mechanical durability.

Graphene, a zero-bandgap semimetal, has been widely explored as a conductive layer in MoS_2-based heterostructures due to its exceptional electrical properties. Its integration with MoS_2 leads to a significant enhancement in charge transfer efficiency, which in turn strengthens the piezoelectric effect. Pan et al. reported that MoS_2/graphene heterostructures exhibit higher piezoelectric voltage output than monolayer MoS_2 alone [28]. The enhanced charge transfer at the MoS_2/graphene interface facilitates efficient carrier separation, reducing charge recombination and increasing the overall polarization under mechanical stress. In experiments using a piezoelectric nanogenerator (PENG) design, the output voltage from the MoS_2/graphene structure was ~2.5 times higher than that from a pure MoS_2 monolayer device. Graphene also provides mechanical reinforcement to MoS_2, improving its flexibility and durability under repeated mechanical loading. MoS_2/graphene heterostructures exhibit improved mechanical robustness, allowing for sustained piezoelectric performance over prolonged use in flexible electronic applications.

4.3.2 Oxide-based interfaces

The integration of MoS_2 with transition metal oxides (TMOs) such as ZnO and TiO_2 has garnered significant interest for enhancing piezoelectric properties, improving mechanical stability, and increasing environmental robustness. These hybrid structures leverage the inherent piezoelectricity of TMOs while benefiting from the unique electronic and mechanical properties of MoS_2. The synergy between MoS_2 and oxide materials enables the development of highly efficient piezoelectric nanogenerators (PENGs), flexible sensors, and energy-harvesting devices.

ZnO is a well-known piezoelectric material with a wurtzite crystal structure that provides strong polarization under mechanical deformation. When combined with MoS_2, the ZnO layer acts as an active component that amplifies the overall piezoelectric response of the heterostructure. Dong et al. demonstrated that ZnO@MoS_2 heterostructures exhibit a ~ 2.0-fold increase in piezoelectric voltage output compared to MoS_2 monolayers alone [42]. This enhancement is attributed to the interfacial charge transfer between ZnO and MoS_2, which facilitates a more efficient separation of piezo-generated carriers. PFM and KPFM studies confirmed that the ZnO layer reinforces the polarization effect in MoS_2 by inducing additional strain at the heterointerface. ZnO nanorod arrays have been integrated with MoS_2 sheets to create vertically aligned heterostructures with superior mechanical flexibility. These structures maintain high piezoelectric output even under repeated mechanical loading cycles, making them suitable for applications in wearable electronics and self-powered sensors.

5. Potential applications of MoS_2 piezoelectric nanomaterials

The unique and robust piezoelectric properties of MoS_2 nanomaterials have paved the way for a wide array of transformative technological applications. Their high mechanical flexibility, coupled with their strong piezoelectric response, makes MoS_2 an ideal candidate for integration into next-generation devices across various fields, including energy harvesting, biomedical monitoring, nanoelectronics, optoelectronics, and environmental remediation. The following sections explore these potential applications in greater detail, backed by recent research advancements and future prospects (**Figure 4**).

5.1 Energy harvesting and self-powered electronics

MoS_2-based piezoelectric nanomaterials are particularly well-suited for energy harvesting applications due to their efficient conversion of mechanical energy into

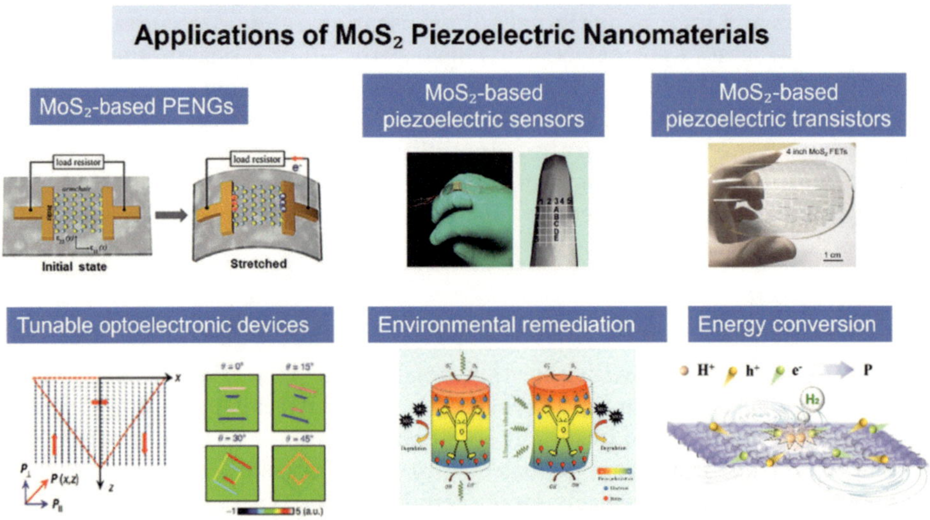

Figure 4.
Potential applications of MoS_2 piezoelectric nanomaterials. Photograph and models of MoS_2-based PENGs, strain sensor, piezoelectric transistors, tunable optoelectronic devices, pollutant degradation, and energy conversion.

electrical power. PENGs fabricated from MoS_2 offer sustainable solutions for self-powered electronics, making them ideal for use in wireless sensors, wearable devices, and Internet of things (IoT) systems.

MoS_2 nanogenerators can convert ambient mechanical energy, such as vibrations, human motion, and wind, into usable electrical energy. Recent studies have demonstrated that MoS_2-based PENGs exhibit excellent power density, with up to 1.4 W/m^2 of output power [43]. When integrated into wearable systems, these devices can power sensors without the need for external batteries, leading to more sustainable and eco-friendly technologies.

The flexibility of MoS_2 allows for the fabrication of stretchable and flexible energy harvesting devices. MoS_2-based PENGs can be integrated into a variety of form factors, including flexible films and textiles, providing new avenues for self-powered smart clothing and flexible electronics [44, 45]. Their scalability also positions MoS_2 as a strong candidate for large-area applications, including smart cities and autonomous systems.

5.2 Biomedical applications: Flexible sensors and monitoring devices

MoS_2's excellent piezoelectric properties, coupled with its biocompatibility and mechanical flexibility, make it an ideal material for real-time physiological monitoring. MoS_2-based piezoelectric sensors can be used for applications such as heartbeat and muscle activity monitoring, enabling advancements in wearable healthcare technology.

The ability of MoS_2 to be integrated into flexible, stretchable sensors allows for continuous, real-time monitoring of vital signs such as heart rate, respiratory rate, and muscle activity. Recent studies have demonstrated that MoS_2-based piezoelectric sensors can detect weak signals, with a high sensitivity of 10 nA/mm^2 [46], making them suitable for wearable bioelectronic devices such as smart patches and health trackers.

The high piezoelectric output of MoS_2 nanomaterials also shows promise in the field of implantable medical devices. MoS_2-based piezoelectric systems could be used for powering implantable sensors or actuators, where mechanical energy from bodily movements is converted into electrical power to recharge batteries or directly drive devices such as neural stimulators [47].

5.3 Nanoelectronics: Flexible logic circuits and memory storage devices

The integration of MoS_2 into nanoelectronic devices offers exciting prospects for the development of next-generation flexible electronics, including logic circuits, memory storage devices, and transistors. The piezoelectric properties of MoS_2 can be exploited in the creation of piezoelectric transistors and diodes, offering new mechanisms for device operation in flexible electronics.

MoS_2-based piezoelectric transistors can modulate charge transport by applying mechanical stress, thus providing a novel approach to electronic switching. Research has shown that MoS_2 field-effect transistors (FETs) can achieve on/off ratios of up to 10^7 under mechanical deformation [48], making them promising candidates for flexible logic circuits. Additionally, MoS_2-based diodes could enable low-power memristor devices for non-volatile memory applications, enhancing the performance and efficiency of flexible memory storage.

5.4 Tunable optoelectronic devices

MoS_2's piezoelectric properties also provide an avenue for the development of tunable optoelectronic devices, where mechanical strain can dynamically modulate the material's optical properties, such as light absorption, emission, and photoluminescence.

MoS_2 nanotubes and nanoribbons, due to their distinct geometries, have demonstrated the ability to tune optical properties under applied mechanical strain. For instance, strain-induced modulation of the bandgap in MoS_2 has been shown to shift its optical absorption and emission wavelengths, offering the potential for reconfigurable photodetectors and light-emitting devices [49]. This could lead to breakthroughs in flexible displays, sensors, and dynamic optical modulators.

The unique combination of MoS_2's piezoelectric and optoelectronic properties can be harnessed in plasmonic devices and photoelectric devices. When combined with plasmonic materials, MoS_2 can enable enhanced light–matter interactions, making it a promising material for use in ultra-sensitive photodetectors and solar cells.

5.5 Piezocatalysis for environmental remediation and energy conversion

The combination of piezoelectricity and catalytic activity in MoS_2 has given rise to a new field known as piezocatalysis. By applying mechanical stress to MoS_2-based materials, the resulting piezoelectric polarization can enhance catalytic reactions, offering potential solutions for environmental remediation and energy conversion.

In piezocatalysis, mechanical strain amplifies catalytic activity by modifying the electronic structure of MoS_2, thereby increasing the efficiency of reactions such as water splitting, carbon dioxide reduction, and pollutant degradation. MoS_2-based piezocatalysts have demonstrated a two-fold increase in catalytic efficiency for water splitting compared to their non-piezoelectric counterparts [50]. This improvement is attributed to the enhanced surface reactivity under applied mechanical stress, which promotes the activation of reaction molecules. MoS_2-based piezocatalysts can also be utilized in energy conversion applications such as hydrogen production via water splitting. The piezoelectric effect facilitates more efficient electron transfer, leading to enhanced hydrogen evolution reaction (HER) rates [51]. This could significantly contribute to the development of green hydrogen production and sustainable energy systems.

6. Conclusion

The exploration of piezoelectric properties in MoS_2 has opened up promising avenues for the design and development of next-generation nanoelectromechanical systems, energy harvesting devices, and multifunctional catalytic platforms. This chapter has provided a detailed overview of the intrinsic and extrinsic factors that govern the piezoelectric behavior of MoS_2, highlighting the critical role of crystal symmetry, strain-induced polarization, and atomic-scale engineering.

Advancements in fabrication techniques—including top-down exfoliation and bottom-up methods such as CVD and ALD—have enabled precise control over MoS_2 thickness, crystallinity, and morphology, which are pivotal for maximizing piezoelectric performance. Furthermore, structural modifications (e.g., nanoribbons and nanosheets), elemental doping (transition metals, non-metals, and vacancy engineering), and interface engineering (with 2D materials and metal oxides) have significantly enhanced the piezoelectric coefficients and expanded the functional versatility of MoS_2.

The integration of MoS_2 piezoelectric nanomaterials into diverse applications—from flexible and wearable electronics to biomedical sensors and piezocatalysts—demonstrates its vast technological potential. Particularly, the coupling of piezoelectricity with catalytic activity in piezocatalysis represents a novel and impactful direction for environmental remediation and sustainable energy conversion.

Looking ahead, key challenges remain in the large-scale production of defect-free, orientation-controlled MoS_2, the long-term stability of devices under cyclic mechanical stress, and the in-depth understanding of piezoelectric coupling mechanisms at the atomic level. Addressing these issues will be essential for the commercial translation of MoS_2-based piezoelectric systems.

Overall, continued interdisciplinary research combining material science, nanotechnology, and device engineering is expected to further unlock the full potential of MoS_2 as a versatile piezoelectric material for advanced applications in the energy, electronics, and environmental sectors.

Acknowledgements

We gratefully acknowledge the financial support from the National Natural Science Foundation of China (Grant No. 22106139) and Zhejiang Provincial Natural Science Foundation of China (Grant No. LZYQ25B060001).

Conflict of interest

The authors declare no conflict of interest.

Author details

Xiuzhen Wei, Jiaming Hu, Liqiao Chen and Meilan Pan*
College of Environment, Zhejiang University of Technology, Hangzhou, P. R. China

*Address all correspondence to: mlpan@zjut.edu.cn

References

[1] Han Q, Han Z, Wang Y, Zhang S, Fang J, Li H, et al. Enhanced photocatalytic hydrogen evolution by piezoelectric effects based on MoSe2/Se-decorated CdS nanowire edge-on heterostructure. Journal of Colloid and Interface Science. 2023;**630**:460-472

[2] Lei R, Gao F, Yuan J, Jiang C, Fu X, Feng W, et al. Free layer-dependent piezoelectricity of oxygen-doped MoS2 for the enhanced piezocatalytic hydrogen evolution from pure water. Applied Surface Science. 2022;**576**:151851

[3] Zhou R, Wu D, Ma J, Ruan L, Feng Y, Ban C, et al. Boosting CO2 piezo-reduction via metal-support interactions in Au/ZnO based catalysts. Journal of Colloid and Interface Science. 2024;**661**:512-519

[4] Martis J, Susarla S, Rayabharam A, Su C, Paule T, Pelz P, et al. Imaging the electron charge density in monolayer MoS2 at the Ångstrom scale. Nature Communications. 2023;**14**:4363

[5] Tsai H-S, Huang Y-H, Tsai P-C, Chen Y-J, Ahn H, Lin S-Y, et al. Ultrafast exciton dynamics in scalable monolayer MoS2 synthesized by metal sulfurization. ACS Omega. 2020;**5**:10725-10730

[6] Wu W, Wang L, Li Y, Zhang F, Lin L, Niu S, et al. Piezoelectricity of single-atomic-layer MoS2 for energy conversion and piezotronics. Nature. 2014;**514**:470-474

[7] Zhu H, Wang Y, Xiao J, Liu M, Xiong S, Wong ZJ, et al. Observation of piezoelectricity in free-standing monolayer MoS2. Nature Nanotechnology. 2015;**10**:151-155

[8] Peng Y, Que M, Tao J, Wang X, Lu J, Hu G, et al. Progress in piezotronic and piezo-phototronic effect of 2D materials. 2D Materials. 2018;**5**:42003

[9] Duerloo K-AN, Ong MT, Reed EJ. Intrinsic piezoelectricity in two-dimensional materials. Journal of Physical Chemistry Letters. 2012;**3**:2871-2876

[10] Wang ZL. Self-powered nanosensors and nanosystems. Advanced Materials. 2012;**24**:280-285

[11] Hinchet R, Khan U, Falconi C, Kim S-W. Piezoelectric properties in two-dimensional materials: Simulations and experiments. Materials Today. 2018;**21**:611-630

[12] Pang F, Zhao P, Lee HY, Kim DJ, Meng X, Cho YS, et al. Progress and perspectives in 2D piezoelectric materials for piezotronics and piezo-phototronics. Advancement of Science. 2025;**2411422**

[13] Kang J, Tongay S, Zhou J, Li J, Wu J. Band offsets and heterostructures of two-dimensional semiconductors. Applied Physics Letters. 2013;**102**:1-21

[14] Wang ZL. Nanopiezotronics. Advanced Materials. 2007;**19**:889-892

[15] Wang X, Feng H, Wu Y, Jiao L. Controlled synthesis of highly crystalline MoS2 flakes by chemical vapor deposition. Journal of the American Chemical Society. 2013;**135**:5304-5307

[16] Li H, Wu J, Yin Z, Zhang H. Preparation and applications of mechanically exfoliated single-layer and multilayer MoS2 and WSe2 nanosheets. Accounts of Chemical Research. 2014;**47**:1067-1075

[17] Mak KF, Lee C, Hone J, Shan J, Heinz TF. Atomically thin MoS2: A new

direct-gap semiconductor. Physical Review Letters. 2010;**105**:136805

[18] Yu H, Zhu H, Dargusch M, Huang Y. A reliable and highly efficient exfoliation method for water-dispersible MoS2 nanosheet. Journal of Colloid and Interface Science. 2018;**514**:642-647

[19] Jurca T, Moody MJ, Henning A, Emery JD, Wang B, Tan JM, et al. Low-temperature atomic layer deposition of MoS2 films. Angewandte Chemie (International Ed. in English). 2017;**56**:4991-4995

[20] Novoselov KS, Mishchenko A, Carvalho A, Castro Neto AH. 2D materials and van der Waals heterostructures. Science. 2016;**353**:9439

[21] Ling X, Wang H, Huang S, Xia F, Dresselhaus MS. The renaissance of black phosphorus. Proceedings of the National Academy of Sciences. 2015;**112**:4523-4530

[22] Lee YH, Zhang XQ, Zhang W, Chang MT, Lin CT, Chang KD, et al. Synthesis of large-area MoS2 atomic layers with chemical vapor deposition. Advanced Materials. 2012;**24**:2320-2325

[23] Liu KK, Zhang W, Lee YH, Lin YC, Chang MT, Su CY, et al. Growth of large-area and highly crystalline MoS2 thin layers on insulating substrates. Nano Letters. 2012;**12**:1538-1544

[24] Bai H, Ding S, Dai Y, Liu J, Chen H, Feng W, et al. Cobalt single-atom intercalation in molybdenum disulfide enhances piezocatalytic and enzyodynamic activities for advanced cancer therapeutics. Advancement of Science. 2025;**12**:2415485

[25] Mondal S, Dilly Rajan K, Patra L, Rathinam M, Ganesh V. Sulfur vacancy-induced enhancement of piezocatalytic H2 production in MoS2. Small. 2025;**21**:2411828

[26] Cysne TP, Guimarães FSM, Canonico LM, Costa M, Rappoport TG, Muniz RB. Orbital magnetoelectric effect in nanoribbons of transition metal dichalcogenides. Physical Review B. 2023;**107**:1-18

[27] Xu X, Liang T, Kong D, Wang B, Zhi L. Strain engineering of two-dimensional materials for advanced electrocatalysts. Materials Today Nano. 2021;**14**:100111

[28] Pan M, Liu S, Chew JW. Unlocking the high redox activity of MoS_2 on dual-doped graphene as a superior piezocatalyst. Nano Energy. 2020;**68**:104366

[29] Fu Y, Ren Z, Wu J, Li Y, Liu W, Li P, et al. Direct Z-scheme heterojunction of ZnO/MoS2 nanoarrays realized by flowing-induced piezoelectric field for enhanced sunlight photocatalytic performances. Applied Catalysis B: Environmental. 2021;**285**:119785

[30] Wang X, Song J, Liu J, Wang ZL. Direct-current Nanogenerator driven by ultrasonic waves. Science. 2007;**316**:102-105

[31] Choi W, Lahiri I, Seelaboyina R, Kang YS. Synthesis of graphene and its applications: A review. Critical Reviews in Solid State and Materials Sciences. 2010;**35**:52-71

[32] Li H, Yin Z, He Q, Li H, Huang X, Lu G, et al. Fabrication of single- and multilayer MoS_2 film-based field-effect transistors for sensing NO at room temperature. Small. 2012;**8**:63-67

[33] Xie L, Jiao L, Dai X, Shao Q, Wang H, Wang Y. Engineering the piezoelectric properties of monolayer MoS_2 via

transition metal doping: A first-principles study. Journal of Physical Chemistry C. 2020;**124**:4717-4725

[34] Wu W, Wang L, Li Y, Zhang F, Lin L, Niu S, et al. Piezoelectricity of single-atomic-layer MoS_2 for energy conversion and piezotronics. Nature. 2014;**514**:470-474

[35] Ma D, Ju W, Li T, Zhang X, He C, Ma B, et al. Modulating electronic, magnetic and chemical properties of MoS2 monolayer sheets by substitutional doping with transition metals. Applied Surface Science. 2016;**364**:181-189

[36] Wang J, Li X, Wang Z, Huang B, Dai Y. Enhancing the piezoelectricity of monolayer MoS_2 by strain and vacancy engineering: A first-principles study. Journal of Physical Chemistry C. 2017;**121**:6704-6710

[37] Chen S, Wu M, Zeng XC. Manipulating piezoelectricity in monolayer MoS_2 via atomic substitution and strain engineering. Applied Physics Letters. 2017;**111**:13101

[38] Ong MT, Reed EJ. Engineered piezoelectricity in graphene. ACS Nano. 2012;**6**:1387-1394

[39] Guan Y, Zhou G, Li L, Jiang Y, Yin J, Liu C, et al. Cu-doped oxygen-rich vacancy MOFs derived perovskite for enhanced mineralization of refractory organics through synergistic non-radical species effects. Separation and Purification Technology. 2024;**335**:126072

[40] Alidoosty-Shahraki M, Bashirpour M. Realistic Modeling of MoS_2 piezoelectric transistor. IEEE Transactions on Electron Devices. 2020;**67**:3459-3465

[41] Lee JH, Park JY, Cho EB, Kim TY, Han SA, Kim TH, et al. Reliable piezoelectricity in bilayer WSe2 for piezoelectric Nanogenerators. Advanced Materials. 2017;**29**:1606667

[42] Dong W, Qian F, Jiayi T, Hongmiao Z, Rui L, Difu Z, et al. Piezoelectric polarization induced by dual piezoelectric materials ZnO nanosheets/MoS2 heterostructure for enhancing photoelectrochemical water splitting. Journal of Colloid and Interface Science. 2024;**653**:1166-1176

[43] Ma R, Zhou G, Gu M, Tang X, Ding W, Guan Y, et al. Cobalt leaching inhibition: Transforming coordination polymers into spherical Co3O4@NC catalysts for accelerated tetracycline degradation via enhanced PMS activation. Applied Surface Science. 2024;**648**:158980

[44] Hedau B, Kang B-C, Ha T-J. Enhanced triboelectric effects of self-poled MoS2-embedded PVDF hybrid nanocomposite films for Bar-printed wearable triboelectric Nanogenerators. ACS Nano. 2022;**16**:18355-18365

[45] Sun C, Zu G, Wei Y, Song X, Yang X. Flexible triboelectric Nanogenerators based on electrospun poly(vinylidene fluoride) with MoS(2)/carbon nanotube composite nanofibers. Langmuir. 2022;**38**:1479-1487

[46] Wu J, Xu Q, Lin E, Yuan B, Qin N, Thatikonda SK, et al. Insights into the role of ferroelectric polarization in Piezocatalysis of Nanocrystalline $BaTiO_3$. ACS Applied Materials & Interfaces. 2018;**10**:17842-17849

[47] Yao Z, Deng J, Li L. Piezoelectric performance regulation from 2D materials to devices. Matter. 2024;7:855-888

[48] Sebastian A, Pendurthi R, Choudhury TH, Redwing JM, Das S.

Benchmarking monolayer MoS2 and WS2 field-effect transistors. Nature Communications. 2021;**12**:693

[49] Kravets VG, Wu F, Auton GH, Yu T, Imaizumi S, Grigorenko AN. Measurements of electrically tunable refractive index of MoS2 monolayer and its usage in optical modulators. npj 2D Materials and Applications. 2019;**3**:36

[50] Ren T, Tian W, Shen Q, Yuan Z, Chen D, Li N, et al. Enhanced piezocatalysis of polymorphic few-layered MoS2 nanosheets by phase engineering. Nano Energy. 2021;**90**:106527

[51] Su Y, Zhang L, Wang W, Li X, Zhang Y, Shao D. Enhanced H2 evolution based on ultrasound-assisted piezo-catalysis of modified MoS2. Journal of Materials Chemistry A. 2018;**6**:11909-11915

Chapter 3

Contact Engineering and Doping Techniques for 2D Semiconductors

Yasir Hassan, Muhammad Sabbtain Abbas and Min Sup Choi

Abstract

This chapter provides an in-depth analysis of advanced contact and doping techniques to optimize the performance of two-dimensional (2D) semiconductors, which are promising candidates for next-generation electronic and optoelectronic devices. Contact engineering techniques, including top and bottom contacts, transferred contacts, van der Waals (vdW), edge contacts, and semi-metallic contacts, are systematically explored. These approaches address critical issues such as Schottky barrier formation, Fermi level pinning, and interface damage caused by conventional deposition techniques. For instance, bottom and transferred contacts minimize interfacial defects, while vdW, edge, and semi-metallic contacts mitigate metal-induced gap states and ensure clean, defect-free interfaces, enhancing charge injection efficiency and device performance. In parallel, the chapter examines doping strategies to modulate the electronic properties of 2D materials. Oxidation doping introduces oxygen atoms to substitute sulfur vacancies, enabling p-type doping with controlled structural integrity. Laser-induced doping leverages precision energy delivery to tailor carrier concentration and reduce contact resistance, while charge transfer doping achieves either n-type or p-type behavior through surface-adsorbed dopants like benzyl viologen, Magic Blue, and metal oxides without compromising the crystal structure. Each method is discussed with its mechanisms, advantages, and challenges, including issues like contamination, scalability, and stability.

Keywords: 2D materials, contact engineering, doping, Schottky barrier, Fermi level pinning, charge injection, transition metal dichalcogenides (TMDs)

1. Introduction

Two-dimensional (2D) layered materials have attracted significant research attention due to their exceptional electronic, optical, and mechanical properties. Transition metal dichalcogenides (TMDs), such as MoS_2, are a diverse class of layered 2D materials that hold great promise in state-of-the-art practical applications, including low contact resistance and high on/off ratio field-effect transistors (FETs), high-performance logic and memory devices, atomically thin flexible bioelectronics, photodetectors with a broad spectral range, and photovoltaic devices with reversible photovoltaic polarity [1–5]. In addition, recent reports have also presented that the intrinsic properties of these TMDs can be tuned *via* external stimuli such as light, stress, temperature, electric and magnetic field for diverse applications [6]. These atomically thin semiconductors-based devices are capable of maximizing the current at low voltage and power operation by reducing the channel length, in contrast to the physical limit in silicon-based devices. However, fully realizing their potential as an alternative to silicon-based devices requires addressing critical challenges, particularly the quantum limits of contact resistance and ballistic transport *via* contact engineering [7]. The electrical contacts formed between metal electrodes and 2D semiconductors significantly impact device performance, making this a crucial factor to consider. Various factors, such as the presence of defects, Schottky barrier height (SBH), and Fermi level pinning, strongly influence charge transport and injection efficiency in electronic devices.

To overcome these challenges, a range of contact engineering techniques has been developed, including top and bottom contacts, transferred contacts, van der Waals (vdW) contacts, edge contacts, and semi-metallic contacts. These contact engineering techniques effectively mitigate metal-induced gap states (MIGS) and facilitate the formation of defect-free, sharp metal–semiconductor interfaces. Additionally, doping engineering has emerged as an effective approach to fine-tuning the electronic properties of 2D materials. Techniques such as oxidation doping, laser-induced doping, and charge transfer doping not only enhance the electrical performance of devices but also enable the tailoring of 2D materials for specific applications. Furthermore, *in-situ* doping methods, including bulk and low-dimensional doping approaches, enable CMOS-compatible doping processes suitable for scalable electronic devices.

This chapter provides a detailed discussion of contact engineering and doping strategies, focusing specifically on MoS_2-based electronic devices, their fabrication processes, and the associated challenges. By addressing issues related to interface quality and contact resistance, these advancements pave the way for the integration of 2D materials into high-performance electronic and optoelectronic platforms.

2. Contact engineering

Contact engineering for 2D semiconductors represents one of the most critical and challenging aspects of the semiconductor industry in the development of high-performance electronic and optoelectronic devices [8]. Device performance is heavily influenced by the quality of the interface between 2D semiconductors and metal contacts [9]. A clean, well-formed interface promotes ohmic contact, enabling efficient electron injection, while the presence of residues or oxide layers at the interface often results in Schottky contact, leading to high contact resistance [10]. To achieve high-performance devices, researchers have developed and reported various techniques tailored to specific performance goals and research objectives. Below, we discuss some of the most commonly employed methods for establishing contacts with 2D materials.

2.1 Conventional top and bottom contacts

Top contacts involve depositing a metal layer directly onto the surface of a 2D material. Typically, the 2D semiconductor is first transferred onto a substrate, after which the metal deposition is performed using methods such as electron beam deposition, thermal evaporation, or sputtering. Prior to deposition, the contact regions are patterned using electron beam lithography (EBL) or photolithography. Commonly used metals for top contacts include gold (Au), chromium (Cr), palladium (Pd), titanium (Ti), nickel (Ni), and silver (Ag). This method is widely utilized in planar device geometries, such as FETs.

Top contact geometry, however, presents several drawbacks that can limit device performance. A major issue is the occurrence of chemical reactions and physical bombardment at the metal–semiconductor interface during metal deposition, which can cause significant damage. This is evident from distorted regions observed in cross-sectional TEM images (**Figure 1a** and **b**) [11]. Another critical challenge is Fermi level pinning, which results in a high SBH and hinders efficient charge injection and transport. The degree of Fermi level pinning is quantified by the pinning factor (S), which depends on both the material and the metal electrode. For example, the pinning factor S is 0.11 for monolayer MoS_2 and 0.07 for monolayer $MoTe_2$ (**Figure 1c**). Due to these interfacial challenges, devices with top contacts using MoS_2 typically exhibit metal-independent n-type behavior and a gradual increase in conduction in their transfer characteristics.

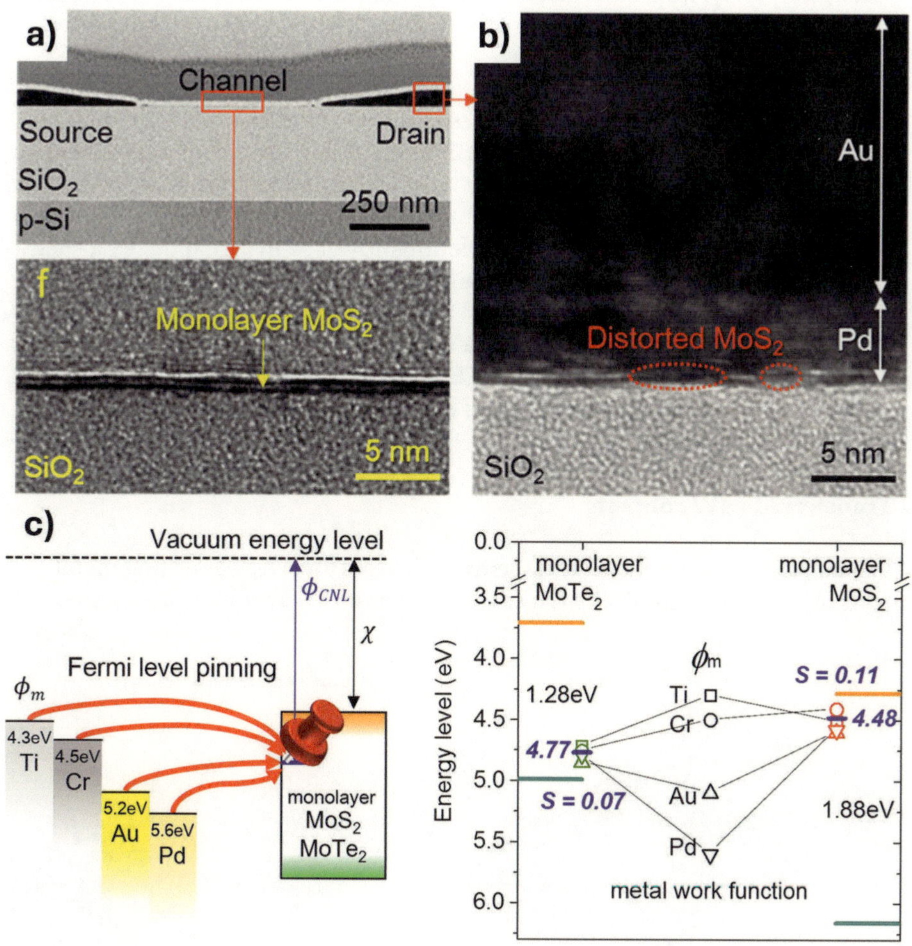

Figure 1.
(a, b) Cross-sectional TEM images of a top-contacted MoS_2 device with Pd electrical contacts, revealing significant structural damage to MoS_2 induced by metal evaporation. (c) Schematic illustration of various metals with different work functions relative to MoS_2, along with their corresponding pinning factors for monolayer MoS_2 and $MoTe_2$. Images reproduced with permission [11]. Copyright 2017 American Chemical Society.

In contrast to top contacts, bottom contact geometry involves depositing metal electrodes onto the substrate first, followed by transferring the 2D material onto the pre-patterned contacts, as illustrated in **Figure 2**. This method minimizes interfacial damage because the 2D material is not exposed to the metal deposition process, resulting in a cleaner, damage-free interface with a suppressed pinning effect.

The cleaner interface provided by bottom contacts leads to a sharp increase in conduction in their transfer characteristics [12]. This improvement originates from a significantly reduced SBH, which facilitates more efficient charge injection Additionally, bottom contact geometry is particularly well-suited for vertical device architectures, such as tunnel junctions and memory devices. However, this approach requires a high degree of precision and expertise to ensure the 2D material is transferred without cracks or structural damage.

A direct comparison of the two approaches highlights their respective strengths and limitations:

- *Interface quality*: Bottom contacts achieve a cleaner interface compared to top contacts, where metal deposition often induces chemical reactions and physical damage.

- *Schottky barrier and conduction*: Devices with top contacts exhibit a gradual increase in conduction due to a larger SBH, whereas bottom contacts enable sharp switching in conduction, attributed to reduced interfacial defects and lower SBH.

- *Device geometry*: Top contacts are predominantly used in planar devices, such as FETs, while bottom contacts are more suitable for vertical architectures.

These distinctions are visually supported by TEM images (**Figure 1a** and **b**) and transfer characteristic comparisons. The TEM images highlight interface damage in top contact devices, with distorted regions encircled to emphasize the detrimental effects on device performance [11, 13].

2.2 Transferred vdW contacts

The fabrication of contacts on 2D semiconductors involves depositing metal contacts onto a clean, atomically flat substrate. These contacts are subsequently

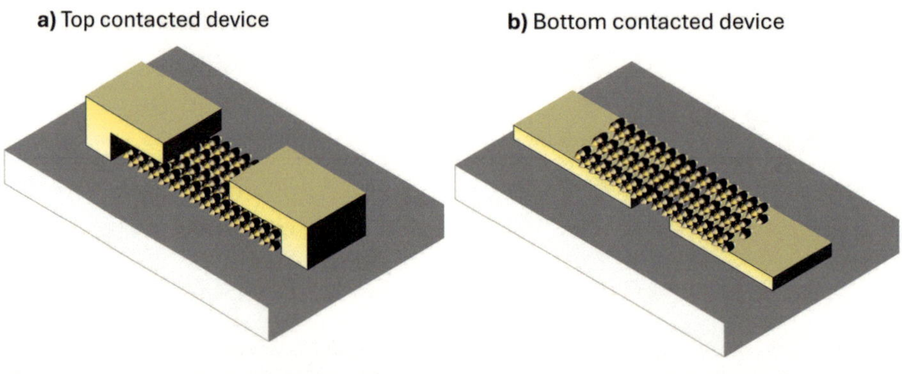

Figure 2.
Device schematics of 2D FETs with (a) top and (b) bottom contact geometries.

transferred onto 2D materials positioned on a separate substrate. This method reduces direct chemical bonding and minimizes physical defects or damage typically associated with direct metal deposition, ensuring a clean and well-defined metal-semiconductor interface.

Liu et al. reported the fabrication process of transferred contacts on MoS_2 [14]. In this method, Au electrodes are initially deposited onto a silicon (Si) substrate. The pre-patterned metal contacts are then mechanically lifted using a polydimethylsiloxane (PDMS) stamp and aligned onto mechanically exfoliated MoS_2 prepared on a PMMA/Si substrate. The transferred electrodes create a clean and uniform metal-semiconductor interface, whereas direct Au deposition onto MoS_2 often induces chemical reactions at the interface, leading to structural damage. The cross-sectional TEM images of the Au/MoS_2 interface displays structural defects and damage that are evident with direct metal deposition, while the transferred electrodes show no such damage. Furthermore, the MoS_2 crystal structure remains intact using the transfer technique, even after electrode removal, whereas direct evaporation causes damage, often leading to the MoS_2 peeling off with the Au electrodes. This damage-free interface facilitates the fabrication of sub-1-nm MoS_2 vertical FETs without a complex lithography process [15].

While this method effectively preserves the integrity of 2D materials, potential contamination from the PDMS stamp should be considered. To mitigate this issue, an alternative approach, known as *via contacts*, involves depositing metal into the vias of the pre-patterned hexagonal boron nitride (hBN) layer [16, 17]. This hBN/metal layer is then lifted using a polypropylene carbonate (PPC)/PDMS stamp, as illustrated in **Figure 3a** [17]. The metal-embedded hBN stack is subsequently transferred onto the 2D semiconductor, ensuring a clean and polymer-free interface (**Figure 3b**). Transfer curves obtained using the via contact method exhibit clear p-type semiconducting behavior with sharp on/off switching, as shown in **Figure 3c**, due to the high-quality metal-semiconductor interface. Park et al. recently demonstrated the application of high work function metallic TMDs, such as TaS_2 and $NbSe_2$, as p-type transferred contacts for high-performance 2D vertical p-FETs [18].

The purpose of transferred contacts is to create vdW interface between metals and 2D semiconductors, thereby preventing the formation of defects during the metallization process. A similar vdW contact interface with 2D semiconductors can also be achieved without the transfer process. These defect-free vdW interfaces can be established by employing thermally decomposable polymers or selenium layers as buffer materials [19, 20]. Such contact methods enable Fermi level depinning with a pinning factor of ~0.9 and facilitate universal contact formation with various metals, including Ag, Al, Ti, Cr, Ni, Cu, Co, Au, and Pd. Due to the defect-free contact interface, the fabricated FET devices exhibited a low contact resistance of 1.25 kΩ-μm and high mobility of 155 $cm^2/V\text{-}s$. The universality and industry fabrication compatible process extend beyond 2D semiconductors to other semiconductor materials, allowing for the exploration of contact properties in next-generation electronic devices.

2.3 1D edge contacts

Edge contact, also known as 1D contact, is an innovative technique for forming metal contacts exclusively at the 1D edge of 2D materials, in contrast to conventional top or bottom contacts. This method offers significant advantages, including Fermi level depinning and thickness-dependent carrier transport across the channel, due to the direct coupling of the metal with the active electronic states of the material,

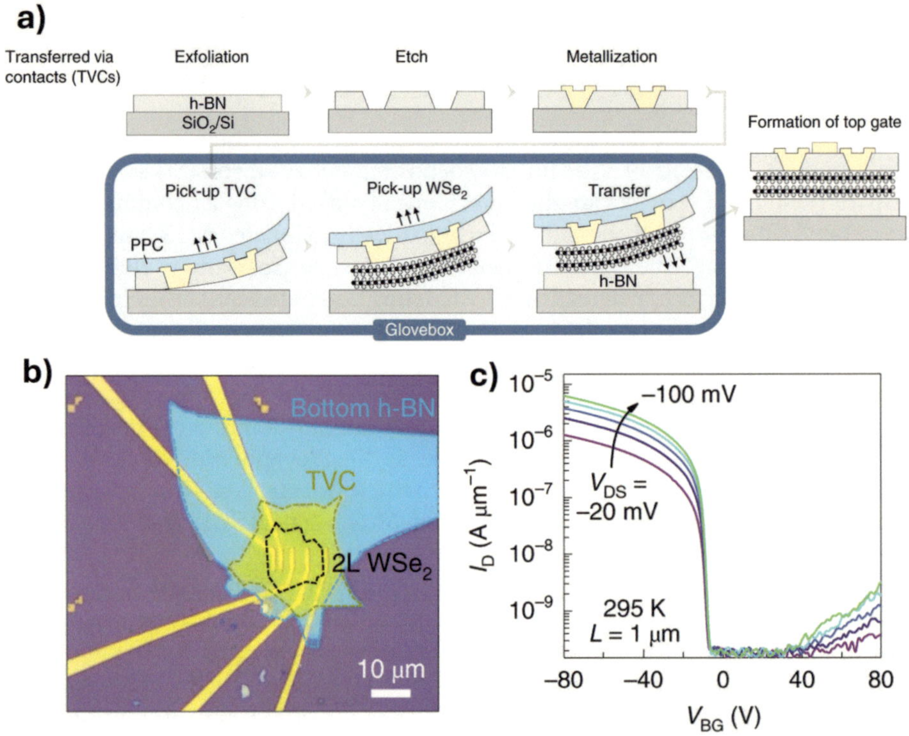

Figure 3.
(a) Schematic illustration of the fabrication process for devices featuring a metal-embedded hBN contact stack. (b) Optical micrograph and (c) transfer curves of WSe_2FETs with transferred via contacts. Images reproduced with permission [17]. Copyright 2019 The Authors. Published by Springer Nature.

as illustrated in **Figure 4a** [21]. As consequently, it significantly reduces the contact resistance, enabling phonon-limited mobility in graphene devices and quantum-limited contact resistance in MoS_2 devices [22, 24, 25].

Edge contacts are typically fabricated through plasma etching using gases such as O_2, CF_4, or SF_6, which ensure a clean and sharp interface prior to metal deposition. This technique is particularly effective in double-gated devices with hBN encapsulation, facilitating the exploration of exotic phenomena such as unconventional superconductivity in twisted graphene and ferroelectricity [26, 27]. The hBN-sandwiched structure preserves the integrity of both the material and its electronic properties. Furthermore, edge contacts hold significant potential in FETs, enabling high mobility, ultra-low contact resistance, and immunity to contact scaling, as shown in **Figure 4b** [21].

Despite its advantages, edge contact fabrication presents several challenges that require further development. One major issue is the instability of the edges in 2D materials, which makes them more susceptible to oxidation or metallization during plasma etching compared to graphene. Another critical challenge is optimizing the plasma etching process, including the selection of appropriate gases and etching durations. Achieving a clean, well-defined interface necessitates precise control to prevent oxidation and ensure proper etching of the heterojunction. Ngo et al. showed a cross-sectional TEM image of an edge contact device fabricated with an hBN/WSe_2/hBN heterostructure, highlighting the oxide layer formed at the contact interface during plasma etching [28].

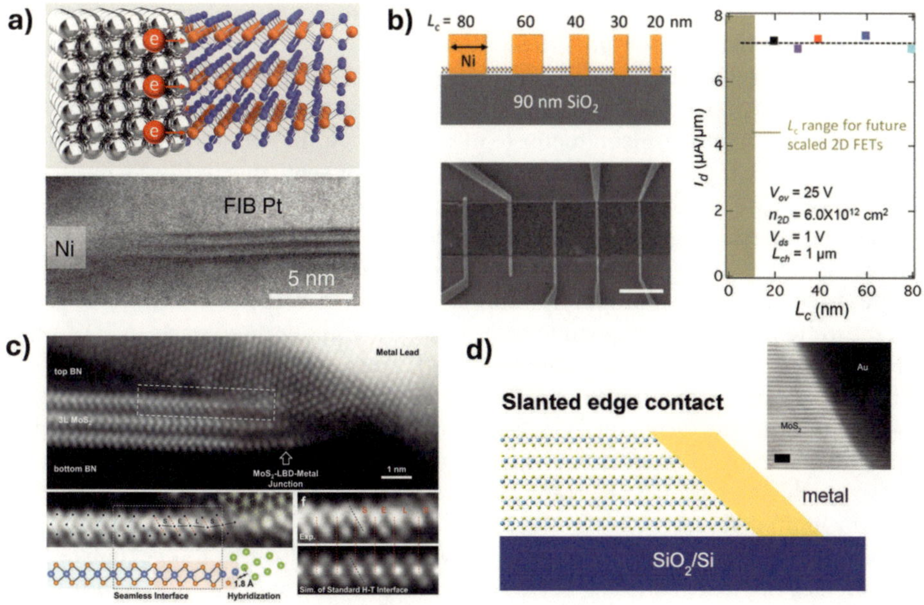

Figure 4.
(a) Schematic of an edge-contacted MoS_2 device, where charge transport between the metal and semiconductor occurs exclusively at the one-dimensional edge interface. (b) Contact scaling effects in one-dimensional edge-contacted MoS_2 FETs. Images reproduced with permission [21]. Copyright 2019, American Chemical Society. (c) Cross-sectional TEM image of the edge contact interface, showing interfacial local bonding distortions. Images reproduced with permission [22]. Copyright 2022, The Authors. Published by Springer Nature. (d) Schematic and TEM image of slanted edge-contacted MoS_2 devices fabricated via a directional angled etching process. Image reproduced with permission [23]. Copyright 2025, The Authors. Published by the American Chemical Society.

Several studies have proposed strategies to address these challenges and achieve sharp, clean 1D edge contact interfaces, as depicted in **Figure 4c** [22]. For example, Cai et al. demonstrated nearly barrier-free contacts by engineering interfacial bonding distortions in MoS_2. Using soft O_2 plasma treatment, they induced local bonding distortions, resulting in a seamless edge interface. The metallic nature of the distorted MoS_2 region led to ultra-low contact resistance, approaching the theoretical quantum limit. In another approach, Lin et al. introduced a slanted edge contact strategy, as shown in **Figure 4d** [23]. By employing directional angled etching, they achieved a well-aligned, sharp 1D metal-semiconductor interface, as evident in the TEM image of **Figure 4d**. This slanted edge structure also provides a larger contact area for elemental analysis of the chemical environment at the interface. Temperature-dependent transport measurements revealed that direct and Fowler-Nordheim (FN) tunneling dominate electron conduction, indicating the formation of a thin tunneling barrier that facilitates efficient carrier injection. Further development of these techniques is essential to achieve high-quality edge contact interfaces.

2.4 Semi-metallic contacts

Contact resistance is a critical factor in the performance of electronic devices, particularly those based on 2D semiconductors. Its primary source is the Schottky barrier formed at the metal-semiconductor interface. This barrier arises mainly from two factors: the work function mismatch between the metal and the semiconductor, and the rehybridization of the semiconductor's electronic states upon contact with the metal, a phenomenon known as MIGS [29].

To address this challenge, various strategies have been proposed, such as doping the semiconductor to reduce the Schottky barrier or introducing a thin dielectric layer between the metal and semiconductor to minimize their interaction. This section focuses on the doping approach, particularly the use of semi-metals like bismuth (Bi) and antimony (Sb) as contact metals for 2D semiconductors [29–31]. The Fermi level of these semi-metals aligns near the semiconductor's conduction band minimum, facilitating a degenerate state in the semiconductor. Additionally, the near-zero density of states in semi-metals significantly suppresses MIGS, thereby contributing to gap-state saturation. These effects enable near-zero SBH and ohmic contacts in 2D devices.

Recent studies have demonstrated the effectiveness of Bi-based ohmic contacts in 2D MoS_2devices, achieving an on/off ratio exceeding 10^7 and linear output curves. However, optimal performance with this approach requires high-quality crystals and precise evaporation techniques to avoid gap-state pinning. Notably, the lowest contact resistance achieved was 123 Ω·μm at a carrier density of 1.5×10^{13} cm^{-2}, approaching quantum-limited values.

Although Bi contacts demonstrate the advantages of semi-metals over conventional metals, the weak coupling between metal and semiconductor states, due to the presence of a vdW gap, limits further reductions in contact resistance. Li et al. achieved a significant reduction in contact resistance by utilizing semi-metallic antimony (Sb($01\bar{1}2$)). The strong hybridization between the energy bands of MoS_2 and Sb($01\bar{1}2$) across the vdW gap enhances charge transfer, resulting in a degenerately doped state in MoS_2, similar to that observed with Bi contacts. The fabricated FETs exhibited clear switching behavior, even with a short channel length of 20 nm, and demonstrated ultra-low contact resistance of less than 100 Ω·μm.

The use of semi-metal contacts not only achieved low contact resistance but also introduced unique features compared to conventional metals, such as the modulation of pseudo-junction resistance through top-gating and temperature-insensitive linearity [30–32]. Additionally, beyond the low work functions of Bi and Sb semi-metals, topological semi-metal niobium phosphide (NbP) has been employed for p-type contacts in WS_2 FETs. The high work function and low density of states of NbP enabled a high hole current of 5.8 μA/μm in sub-2-nm-thick WS_2 FETs [33].

3. Doping engineering

Contact engineering is a critical factor in enhancing the performance of devices based on 2D semiconductors. Similarly, doping has emerged as a successful strategy for improving the electrical and optical properties of 2D material-based devices, enabling their use in a wide range of applications. Various doping methods have been explored and implemented in next-generation electronics, as discussed in the following section.

3.1 *Ex-situ* doping

3.1.1 Oxidation doping

Oxidation doping is one of the simplest and most effective techniques for doping 2D semiconductors such as MoS_2 and WSe_2. In this method, oxygen atoms substitute for sulfur (or selenium) vacancies through various techniques, including oxygen plasma treatment, UV-ozone (UVO) exposure, calcination, or heat treatment in an ambient atmosphere [34, 35]. This substitution, particularly *via* oxygen plasma or UVO treatment, introduces defects in the crystal structure, significantly altering the material's electronic and optical properties.

Upon oxidation, 2D semiconductors exhibit p-type doping. However, it is crucial to precisely control the extent of oxidation, as excessive exposure to oxygen plasma or UV treatment can degrade the crystal structure, transforming MoS_2 or WSe_2 into MoO_2 and WO_x compounds. Recent studies have shown that baking MoS_2 at temperatures below 300°C in an ambient atmosphere yields optimal oxidation [36]. Beyond 300°C, excessive oxygen exposure results in structural distortion and decomposition, while baking at 200°C causes a positive threshold voltage to shift in the transfer curves.

Similarly, controlled UVO treatment of MoS_2 and WSe_2 induces p-type doping due to oxygen substitution [37, 38]. UVO treatment offers advantages over oxygen plasma doping, as it involves non-directional chemical reactions rather than physical ion bombardment. Using this approach, a recent study demonstrated the fabrication of a high-performance double-side-doped (DSD) 2D FET channel through substrate hole patterning (**Figure 5a**). The transfer characteristics of the DSD device revealed a stronger p-doped state compared to single-side-doped (SSD) 2D FETs. The enhanced p-doping effects were further confirmed by the degree of redshift observed in the photoluminescence (PL) spectra of SSD and DSD devices (**Figure 5b**). Additionally, MoO_2/WO_x layers have been shown to function as protective coatings, inducing p-doping in graphene-based devices and enabling high carrier mobility [37, 39].

3.1.2 Laser-induced doping

Laser-induced doping is a highly effective technique for defect engineering of 2D TMDs. By irradiating TMDs with a laser, this method enables not only precise patterning but also modulation of carrier concentration, reduction of contact resistance, and surface functionalization [40–42]. Several spectroscopic techniques, including PL and Raman spectroscopy, can be employed to monitor the effects of laser treatment on TMD materials, which directly influence their electrical and optical properties.

The success of laser-induced doping depends on selecting appropriate laser energy and exposure time to achieve optimal adsorption between intrinsic electrons and laser-induced functional groups. However, excessive energy or prolonged exposure can lead to structural deformation or degradation of MoS_2, effects similar to those observed in oxidation doping, as previously discussed. Recent studies have shown that high-power laser irradiation of MoS_2 results in the formation of MoO_x on the surface, while low-power irradiation induces thinning of the MoS_2 flake (**Figure 6**) [41].

Additionally, femtosecond laser irradiation has been demonstrated to reduce contact resistance in MoS_2 devices transferred onto Au electrodes [43]. This process facilitates the adsorption of Au atoms onto the MoS_2 channel, reducing the Schottky barrier at the metal-semiconductor interface and enhancing device performance. Raman spectroscopy provides further insights into laser-induced modifications: a blue shift in the out-of-plane A_{1g} peak indicates p-type doping, whereas a redshift in the in-plane E^1_{2g} peak suggests potential structural distortion [40].

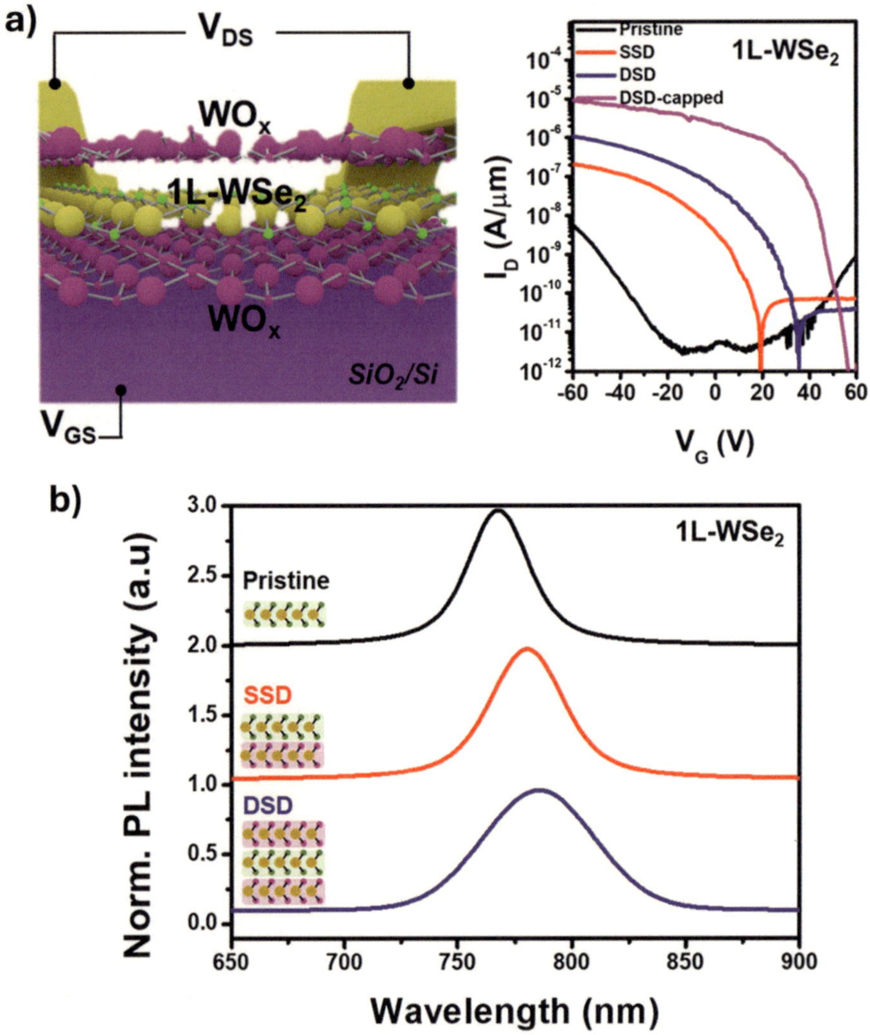

Figure 5.
(a) Schematic and transfer characteristics of a double-side-doped WSe_2 FET treated with UVO, demonstrating p-type doping. (b) PL spectra of pristine, SSD, and DSD WSe_2. Images reproduced with permission [38]. Copyright 2024, American Chemical Society.

3.1.3 Charge transfer doping

Charge transfer doping is an effective method for doping 2D materials, such as MoS_2 and WSe_2, where the exchange of charges between the dopant and the semiconductor modifies the material's electrical and optical properties [44]. In this approach, dopant species physically adsorb onto the surface of the 2D material, facilitating charge transfer that causes the semiconductor channel to act as either an electron donor or acceptor. Unlike other doping mechanisms, charge transfer doping does not introduce external defects into the crystal structure, thereby preserving the intrinsic characteristics of the material [45].

Various organic dopants, such as benzyl viologen (BV), metal oxide dopants like AlO_x/TiO_x, and the rare-earth element yttrium (Y), have been explored for n-type

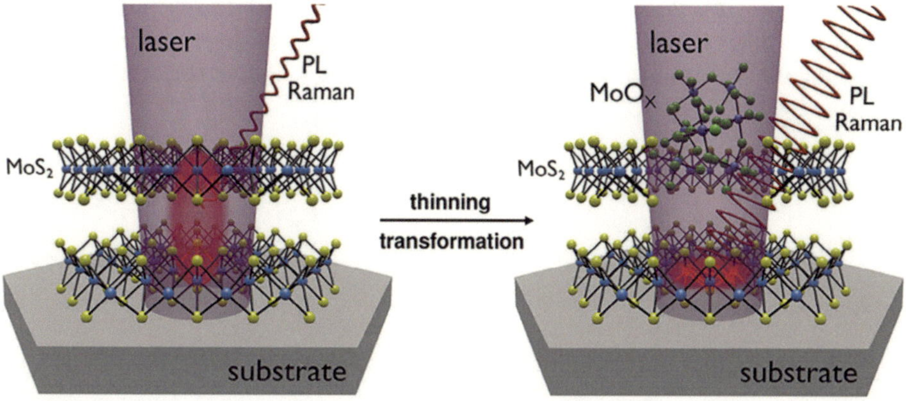

Figure 6.
Schematic illustration of the MoS_2 irradiated by laser light. Images reproduced with permission [41]. Copyright 2020, American Chemical Society.

doping of 2D materials [44–47]. Similarly, tris(4-bromophenyl)ammoniumyl hexachloroantimonate (commonly referred to as "Magic Blue," MB) and MoO_x/WO_x are recognized for p-type doping of 2D materials [48, 49]. For example, in the case of BV, two electrons are transferred to MoS_2, leading to the formation of BV^{2+} molecules and shifting the Fermi level of MoS_2 toward the conduction band, thereby enhancing conductivity at positive voltages (**Figure 7a**) [44]. In yttrium doping, yttrium atoms replace sulfur atoms at the MoS_2 interface, forming a stable metallic MoS_2 structure that enables uniform charge transfer across the channel. Conversely, MB induces the transfer of one electron from MoS_2, creating a hole vacancy and shifting the Fermi level toward the valence band, resulting in p-type doping of the 2D material.

Despite its potential, charge transfer doping faces several challenges, including the air stability of dopants, as many are unstable in ambient environments and must be handled inside a glove box. Additionally, optimizing the choice of dopant and its molar concentration is critical, as certain dopants may lead to degenerate doping, while non-degenerate doping is typically preferred in semiconductor applications. The transfer characteristics of MoS_2 FET devices before and after treatment with various organic dopants at different concentrations reveal that the electrical properties of MoS_2 can vary significantly depending on the type and concentration of n-type and p-type dopants, making it challenging to achieve optimized and stable conditions for 2D FETs [48]. In contrast, **Figure 8b** and **c** illustrate the schematic structure of a MoS_2 channel with AlO_x serving as both a doping and passivation layer [46]. The transfer characteristics of the device demonstrate the stable doping effects achieved with AlO_x, highlighting its potential for reliable performance in 2D devices.

3.2 *In-situ* doping

In-situ doping involves the integration of doping elements into the crystal lattice of a host material during its synthesis, whether in atomic layers or bulk form. Pristine, high-quality MoS_2 layers or flakes can be synthesized using well-established growth techniques, such as chemical vapor deposition (CVD). However, synthesizing *in-situ* doped MoS_2 requires careful consideration of the spatial distribution of doping elements within its atomic structure, either as interstitial or substitutional dopants.

When the dopants are uniformly incorporated throughout the entire atomic structure, the process is referred to as bulk doping, where the dopants form chemical bonds along the z-axis of the bulk MoS_2 crystals or flakes. Conversely, if the dopants

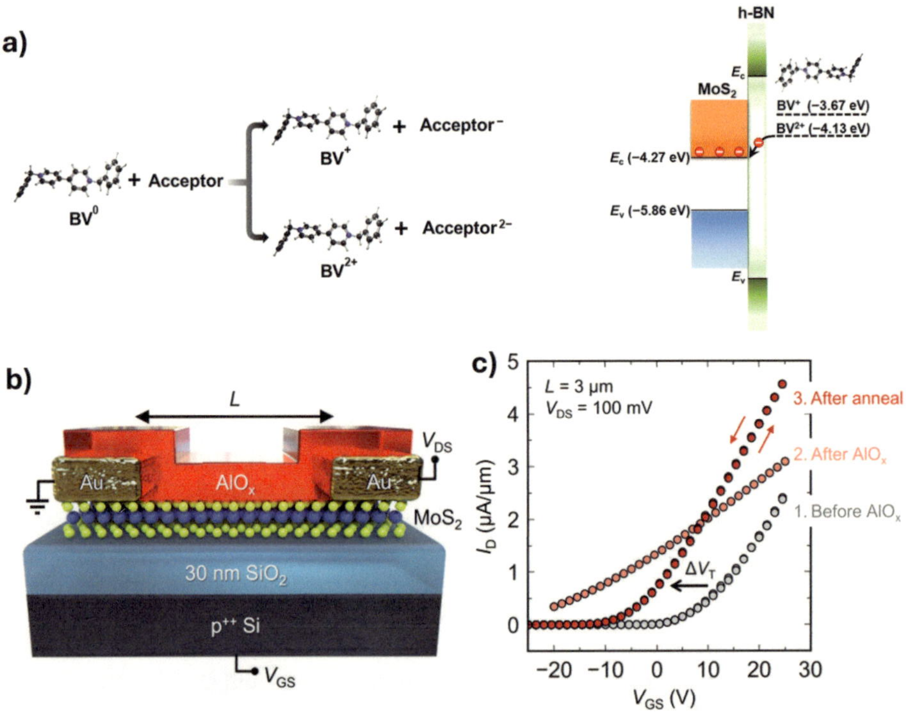

Figure 7.
(a) Schematic of the doping process of MoS_2 illustrating the charge transfer process of BV as an n-dopant. Images reproduced with permission [44]. Copyright 2022, The Authors. Published by the American Association for the Advancement of Science. (b, c) Schematic of a MoS_2 device with a metal oxide layer (AlO_x) and its electrical performance before and after capping with AlO_x. Images reproduced with permission [46]. Copyright 2021, American Chemical Society.

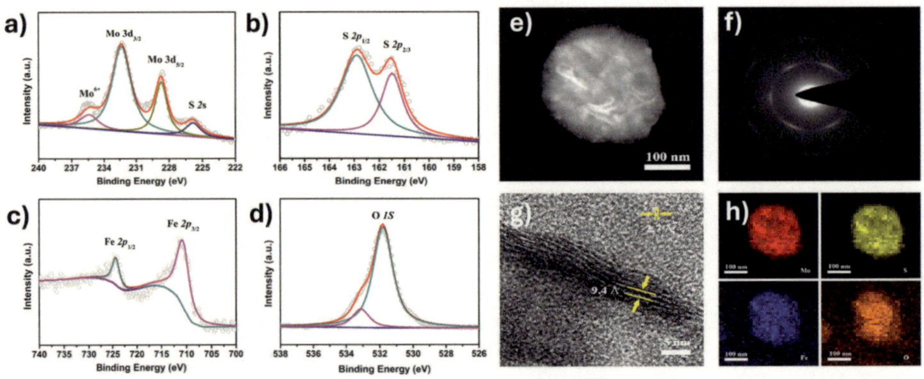

Figure 8.
(a–d) XPS spectra of Mo 3d, S 2p, Fe 2p, and O 1 s for in-situ *iron-doped MoS_2 nanosheets. (e–g) HRTEM images and selected area electron diffraction (SAED) pattern for iron-doped MoS_2. (h) EDS elemental mapping of in-situ iron-doped MoS_2 nanosheets. Images reproduced with permission [50]. Copyright 2019, American Chemical Society.*

are confined to a localized region, such as specific surfaces or edges of the atomic layers, the process is termed low-dimensional doping. These approaches depend on the dimensionality and distribution of the dopants to achieve the desired modifications in MoS_2 properties and align with specific applications.

While doping processes can also be performed *ex-situ*—through techniques such as UVO exposure, plasma oxidation, laser treatment, chemical functionalization, or charge transfer doping, as discussed earlier in this chapter—this section focuses exclusively on *in-situ* doping. A comparative summary of the key features of these mechanisms is provided in **Table 1**.

3.2.1 Bulk in-situ doping

The incorporation of dopants into the lattice structure of bulk MoS_2 crystals can occur either through the substitution of Mo or S atoms within the atomic layers or as interstitials in the vdW gaps between layers. The type of doping element (e.g., transition or alkali metals), its atomic radius, and its concentration determine whether the dopants introduce defects or repair existing ones, induce extra charge carriers (electrons or holes), modify the band structure, or cause lattice strain, resulting in structural, electronic, or mechanical property modifications [51]. This approach is particularly employed for tailoring the properties of MoS_2 crystals at large scales for applications in energy, catalysis, electronics, and optoelectronics.

Solvothermal synthesis is commonly used for bulk *in-situ* doping of MoS_2, where dopants are added to the precursor solution to achieve desired functionalities. **Figure 8a–h** illustrate the spectroscopic and morphological evolution of Fe-doped MoS_2 flakes, designed for efficient hydrogen evolution reactions [50]. The presence of iron in the host MoS_2 structure was confirmed by XPS analysis of key elements (Mo, S, O, and Fe), while electron microscopy images showed uniform, hinged flakes of Fe-doped MoS_2. High-resolution electron microscopy (HRTEM) imaging further revealed an increase in the interlayer spacing to 0.94 nm, compared to the pristine value of 0.62 nm [52]. Additionally, energy dispersive spectroscopy (EDS) elemental mapping (**Figure 8h**) demonstrated uniform incorporation of Fe into the MoS_2 atomic lattice.

Qin et al. employed a cost-effective sol-gel method to incorporate nitrogen into MoS_2 nanosheets, focusing on inducing porosity, curling, and overlapping of the nanosheets for energy storage applications [53]. The Mo 3p and N 1 s XPS spectra

Features	Bulk doping	Low-dimensional doping
Spatial Extent	Along z-direction in bulk flakes	Confined to particular surfaces and edges of mono, bi, and few layers.
Synthesis Techniques	Solution-processed doping (Hydrothermal, Solvothermal, etc.)	CVD, MBE for *in-situ* & Laser and plasma exposure for *ex-situ*
Material Type	Multilayers and Bulk crystal	Targeted or complete coverage of mono, bi, and few layers.
Effect on Properties	Modification in bulk properties	Localized change in properties.

Table 1.
Key features of bulk and low-dimensional in-situ *doping of* MoS_2.

confirm the formation of Mo–N bonds. HRTEM images reveal a sponge-like porous framework formed by curled and overlapped aggregates of nanosheets. Similarly, other metal or non-metal elements can be introduced *via* solvothermal synthesis to finely tune the structure and morphology of MoS_2 for diverse applications.

3.2.2 Low-dimensional in-situ doping

Low-dimensional *in-situ* doping involves the selective addition of dopant elements to specific areas, edges, or surfaces of MoS_2 atomic layers during synthesis. CVD is the most widely used method for synthesizing high-quality pristine and *in-situ* doped MoS_2 [54]. This process involves introducing dopant-containing precursors into the reaction chamber alongside the base precursors of molybdenum (e.g., MoO_3 or $MoCl_5$) and sulfur (e.g., sulfur powder or H_2S gas) [55]. The method enables uniform distribution of dopants in localized regions or across entire atomic layers, imparting multifunctional properties such as monolayer electronics, heterostructure devices, sensors, quantum effects, and single-photon emitters. The quality of doped MoS_2 depends on several factors, including the choice of precursor phase, doping procedure, CVD configuration, dopant concentration, and growth parameters. **Figure 9** outlines various strategies and their key features commonly employed for in-situ low-dimensional doping of MoS_2.

In typical CVD processes, a gas-phase precursor (e.g., O_2, N_2, NH_3) is introduced into the reaction chamber alongside Mo and S source precursors. By precisely controlling growth conditions—particularly temperature and gas concentration—the physical properties of the material can be tuned. Tang et al. investigated the structural, morphological, and FET device performance of *in-situ* oxygen-doped MoS_2 monolayers at varying gas concentrations [56]. XPS spectra confirmed the formation of $MoS_{2-x}O_x$ with Mo–O and Mo–S bonds, and qualitative estimation of oxygen doping levels (ranging from 0 to 26%) was determined by analyzing the areas of these XPS peaks, as shown in **Figure 10a**.

Raman spectra revealed a peak centered at 287 cm^{-1} corresponding to the vibrational mode of Mo–O bonds, while PL spectra showed a decrease in the optical bandgap to 1.8 eV, along with homogeneous doping across $MoS_{2-x}O_x$ triangles in the PL map (**Figure 10b–d**). AFM imaging (**Figure 10e**) indicated no significant height difference (~0.7 nm) in the doped triangles, while high-resolution scanning tunneling microscopy (HRSTM) images (**Figure 10f–g**) showed dark spots representing oxygen substitution, validating the spatial uniformity of the in-situ O_2 doping.

The FET device performance revealed that sheet conductance and mobility can be optimized by carefully controlling the doping level (up to 10%). However, heavy doping drastically reduces carrier mobility due to defect scattering and damage on the

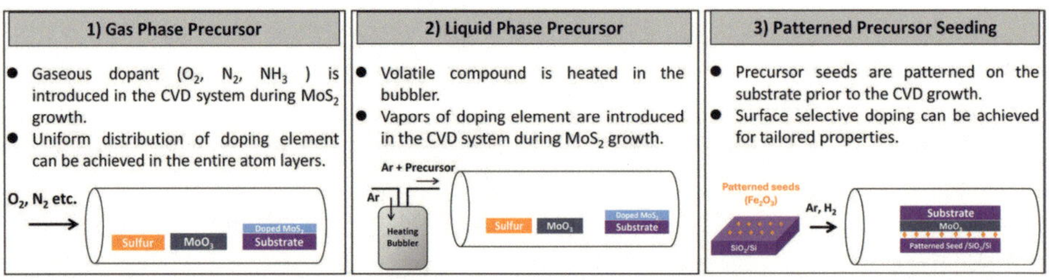

Figure 9.
CVD growth strategies for low-dimensional in-situ *doping in MoS_2.*

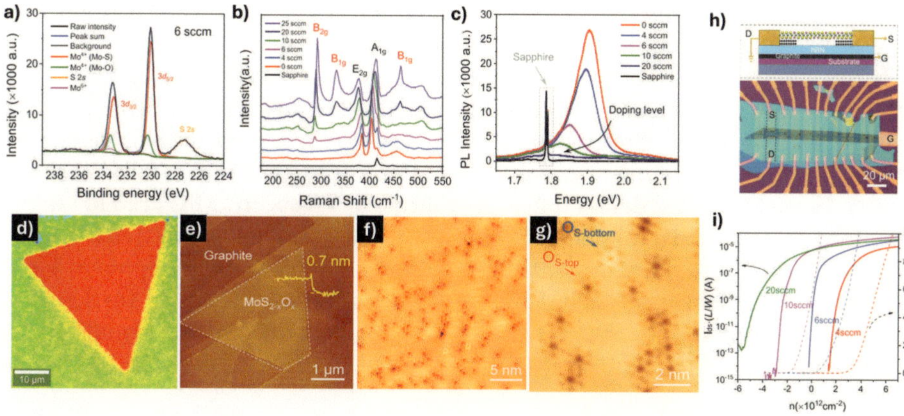

Figure 10.
(a–c) XPS, Raman, and PL spectra of in-situ *oxygen-doped MoS_2 monolayer. (d) PL mapping and (e, f) HR-STM images of an* in-situ *oxygen-doped MoS_2 monolayer. (h) Schematic and optical micrograph of an $MoS_{2-x}O_x$ FET. (i) Normalized transfer curves as a function of carrier density for different oxygen doping levels. Images reproduced with permission [56]. Copyright 2020, The Authors. Published by John Wiley and Sons.*

basal plane of $MoS_{2-x}O_x$, as shown in **Figure 11h–i**. For volatile precursors (e.g., transition metals, chalcogens, or nanomaterials), a heated bubbler is typically attached to the CVD system to transport dopant vapors with the carrier gas, enabling the synthesis of controllable and homogeneous in-situ doped MoS_2, as illustrated in **Figure 9**.

Pre-patterning precursor-containing seeds on the substrate prior to CVD synthesis is another widely used strategy for *in-situ* doping of MoS_2. This method enables surface-selective doping while preserving undoped regions in their intrinsic form, allowing for the fabrication of device-specific structures with tailored properties, such as heterostructures, FETs, and p-n junctions. Fu et al. demonstrated the synthesis of iron (Fe)-doped MoS_2 by casting a patterned seed of a Fe-containing precursor

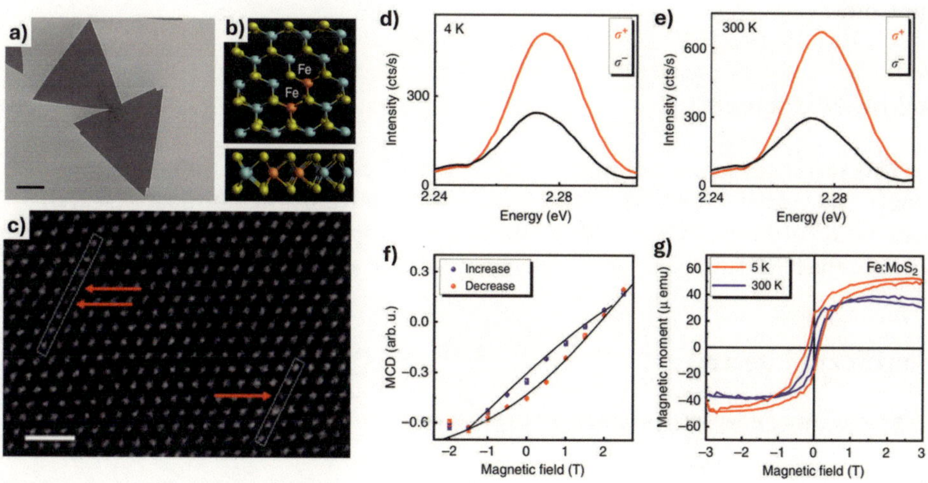

Figure 11.
(a) SEM image, (b) Schematics for top and side views, and (c) STEM image of Fe-MoS_2 monolayer. (d, e) Magneto-photoluminescence measurements and (f, g) MCD and magnetization measurements of a Fe-MoS_2 monolayer at cryogenic and room temperatures. Image reproduced with permission [57]. Copyright 2020, The Authors. Published by Springer Nature.

(Fe_2O_3) onto a SiO_2/Si substrate. This was achieved using a sandwiched configuration, where another MoO_3-coated substrate was placed upside down [57].

The scanning electron microscopy (SEM) image (**Figure 11a**) revealed triangular island-like morphologies, while STEM imaging (**Figure 11c**) confirmed Fe substitution at Mo atomic sites, as illustrated in the schematic top and side views of Fe-MoS_2 (**Figure 11b**). Temperature-dependent PL spectra exhibited emission peaks associated with Fe-related defect states. Furthermore, PL spectra (**Figure 11d** and **e**) obtained under circularly polarized light at 4 K and 300 K demonstrated strong circular dichroism, providing insights into magnetism induced by Fe doping in MoS_2. Magnetic circular dichroism (MCD) and M-H curves (**Figure 11f** and **g**) further confirmed pronounced ferromagnetism in Fe-MoS_2 monolayers at 300 K and 5 K, respectively.

4. Conclusions

This chapter reviewed the state-of-the-art approaches in contact and doping engineering for 2D semiconductors, emphasizing their pivotal role in enhancing electronic and optoelectronic device performance. Contact engineering methods, including top, bottom, transferred vdW, edge, and, semi-metallic contacts, address challenges like interface quality, Fermi level pinning, and contact resistance, with ultraclean and flat contact interfaces emerging as promising solutions for ultra-low resistance and structural preservation. Doping strategies, such as oxidation, laser-induced, and charge transfer doping, enable precise tuning of carrier concentrations while maintaining material integrity. The interplay of these techniques highlights their combined importance in overcoming the limitations of 2D materials and achieving breakthroughs in mobility, contact resistance, and device efficiency. Despite significant advancements, challenges in scalability, defect-free interfaces, and doping precision remain. Addressing these barriers could unlock the full potential of 2D materials for next-generation applications in transistors, photodetectors, and flexible electronics.

Acknowledgements

This work was supported by the Basic Science Research Program (2021R1A6A1A03043682, 2022R1C1C2005607), funded by the National Research Foundation of Korea (NRF). This work was also supported by the Brain Pool program through the NRF funded by the Ministry of Science and ICT (RS-2024-00407437).

Conflict of interest

The authors declares no conflict of interest.

DOI: http://dx.doi.org/10.5772/intechopen.1009620

Author details

Yasir Hassan[1], Muhammad Sabbtain Abbas[2] and Min Sup Choi[1]*

1 Department of Materials Science Engineering, Chungnam National University, Daejeon, South Korea

2 Institute of Applied Physics, Seoul National University, Seoul, South Korea

*Address all correspondence to: goodcms@cnu.ac.kr

References

[1] Mondal A, Biswas C, Park S, Cha W, Kang S-H, Yoon M, et al. Low ohmic contact resistance and high on/off ratio in transition metal dichalcogenides field-effect transistors via residue-free transfer. Nature Nanotechnology. 2024;**19**:34-43. DOI: 10.1038/s41565-023-01497-x

[2] Hong S, Park J, Lee JJ, Lee S, Yun K, Yoo H, et al. Multifunctional molybdenum disulfide flash memory using a PEDOT: PSS floating gate. NPG Asia Materials. 2021;**13**:38. DOI: 10.1038/s41427-021-00307-x

[3] Kireev D, Kutagulla S, Hong J, Wilson MN, Ramezani M, Kuzum D, et al. Atomically thin bioelectronics. Nature Reviews Materials. 2024;**9**:906-922. DOI: 10.1038/s41578-024-00728-4

[4] Chen X-Y, Su D, Li K-H, Song Y-J, Xia P, Zhang X-Y. Multilayer MoS_2 photodetector with broad spectral range and multiband response. Advanced Devices & Instrumentation. 2024;**5**:0042. DOI: 10.34133/adi.0042

[5] Abnavi A, Ahmadi R, Ghanbari H, Akinwande D, Adachi MM. Switchable photovoltaic effect induced by light intensity. ACS Nano. 2024;**18**(50):34147-34157. DOI: 10.1021/acsnano.4c10392

[6] Torres-Cavanillas R, Forment-Aliaga A. Design of stimuli-responsive transition metal dichalcogenides. Communications Chemistry. 2024;**7**:241. DOI: 10.1038/s42004-024-01322-z

[7] Akinwande D, Biswas C, Jena D. The quantum limits of contact resistance and ballistic transport in 2D transistors. Nature Electronics. 2025;**8**:96-98. DOI: 10.1038/s41928-024-01335-5

[8] Shim J, Park HY, Kang DH, Kim JO, Jo SH, Park Y, et al. Electronic and optoelectronic devices based on two-dimensional materials: From fabrication to application. Advanced Electronic Materials. 2017;**3**(4):1600364. DOI: 10.1002/aelm.201600364

[9] Zhang S, Le ST, Richter CA, Hacker CA. Improved contacts to p-type MoS_2 transistors by charge-transfer doping and contact engineering. Applied Physics Letter. 2019;**115**(7):073106. DOI: 10.1063/1.5100154

[10] Park JY, Joe HE, Yoon HS, Yoo S, Kim T, Kang K, et al. Contact effect of ReS_2/metal interface. ACS Applied Materials & Interfaces. 2017;**9**(31):26325-26332. DOI: 10.1021/acsami.7b06432

[11] Kim C, Moon I, Lee D, Choi MS, Ahmed F, Nam S, et al. Fermi level pinning at electrical metal contacts of monolayer molybdenum dichalcogenides. ACS Nano. 2017;**11**(2):1588-1596. DOI: 10.1021/acsnano.6b07159

[12] Li Y, Zhang X, Duan X, Niu W, Zhao S, He X, et al. Electrode engineering in MoS_2 MOSFET: Different semiconductor/metal interfaces. Advanced Electronic Materials. 2022;**8**(10):2200513. DOI: 10.1002/aelm.202200513

[13] Liu X, Choi MS, Hwang E, Yoo WJ, Sun J. Fermi level pinning dependent 2D semiconductor devices: Challenges and prospects. Advanced Materials. 2022;**34**(15):2108425. DOI: 10.1002/adma.202108425

[14] Liu Y, Guo J, Zhu E, Liao L, Lee S-J, Ding M, et al. Approaching the Schottky–Mott limit in van der Waals metal–semiconductor junctions. Nature.

2018;**557**:696-700. DOI: 10.1038/s41586-018-0129-8

[15] Liu L, Kong L, Li Q, He C, Ren L, Tao Q, et al. Transferred van der Waals metal electrodes for sub-1-nm MoS_2 vertical transistors. Nature Electronics. 2021;**4**:342-347. DOI: 10.1038/s41928-021-00566-0

[16] Telford EJ, Benyamini A, Rhodes D, Wang D, Jung Y, Zangiabadi A, et al. Via method for lithography free contact and preservation of 2D materials. Nano Letters. 2018;**18**(2):1416-1420. DOI: 10.1021/acs.nanolett.7b05161

[17] Jung Y, Choi MS, Nipane A, Borah A, Kim B, Zangiabadi A, et al. Transferred via contacts as a platform for ideal two-dimensional transistors. Nature Electronics. 2019;**2**(5):187-194. DOI: 10.1038/s41928-019-0245-y

[18] Park H, Shin H, Ali N, Choi H, Kim BSY, Kang B, et al. P-type vertical FETs realized by using Fermi-level pinning-free 2D metallic electrodes. Nano Letters. 2025;**25**(1):368-375. DOI: 10.1021/acs.nanolett.4c05136

[19] Kong L, Wu R, Chen Y, Huangfu Y, Liu L, Li W, et al. Wafer-scale and universal van der Waals metal semiconductor contact. Nature Communications. 2023;**14**:1014. DOI: 10.1038/s41467-023-36715-6

[20] Kwon G, Choi Y-H, Lee H, Kim H-S, Jeong J, Jeong K, et al. Interaction- and defect-free van der Waals contacts between metals and two-dimensional semiconductors. Nature Electronics. 2022;**5**:241-247. DOI: 10.1038/s41928-022-00746-6

[21] Cheng Z, Yu Y, Singh S, Price K, Noyce SG, Lin Y-C, et al. Immunity to contact scaling in MoS_2 transistors using in situ edge contacts. Nano Letters. 2019;**19**(8):5077-5085. DOI: 10.1021/acs.nanolett.9b01355

[22] Cai X, Wu Z, Han X, Chen Y, Xu S, Lin J, et al. Bridging the gap between atomically thin semiconductors and metal leads. Nature Communications. 2022;**13**:1777. DOI: 10.1038/s41467-022-29449-4

[23] Lin C-C, Paylaga NT, Yen C-C, Lin Y-H, Wang K-H, Watanabe K, et al. Interfacial elemental analysis of slanted edge-contacted monolayer MoS_2 transistors via directionally angled etching. ACS Nano. 2024;**19**:4452-4461. DOI: 10.1021/acsnano.4c13581

[24] Choi MS, Ali N, Ngo TD, Choi H, Oh B, Yang H, et al. Recent progress in 1D contacts for 2D-material-based devices. Advanced Materials. 2022;**34**(39):2202408. DOI: 10.1002/adma.202202408

[25] Wang L, Meric I, Huang PY, Gao Q, Tran H, Taniguchi T, et al. One-dimensional electrical contact to a two-dimensional material. Science. 2013;**342**(6158):614-617. DOI: 10.1126/science.1244358

[26] Cao Y, Fatemi V, Fang S, Watanabe K, Taniguchi T, Kaxiras E, et al. Unconventional superconductivity in magic-angle graphene superlattices. Nature. 2018;**556**:43-50. DOI: 10.1038/nature26160

[27] Weston A, Castanon EG, Enaldiev V, Ferreira F, Bhattacharjee S, Xu S, et al. Interfacial ferroelectricity in marginally twisted 2D semiconductors. Nature Nanotechnology. 2022;**17**:390-395. DOI: 10.1038/s41565-022-01072-w

[28] Ngo TD, Choi MS, Lee M, Ali F, Yoo WJ. Anomalously persistent p-type behavior of WSe_2 field-effect transistors by oxidized edge-induced

Fermi-level pinning. Journal of Materials Chemistry C. 2022;**10**:846-853. DOI: 10.1039/D1TC04148G

[29] Shen PC, Su C, Lin Y, Chou AS, Cheng CC, Park JH, et al. Ultralow contact resistance between semimetal and monolayer semiconductors. Nature. 2021;**593**:211-217. DOI: 10.1038/s41586-021-03472-9

[30] Lee S, Wang X, Shin H, Ali N, Ngo TD, Hwang E, et al. Semi-metal edge contact for barrier-free carrier transport in MoS_2 field effect transistors. ACS Applied Electronic Materials. 2024;**6**(6):4149-4158. DOI: 10.1021/acsaelm.4c00250

[31] Li W, Gong X, Yu Z, Ma L, Sun W, Gao S, et al. Approaching the quantum limit in two-dimensional semiconductor contacts. Nature. 2023;**613**:274-279. DOI: 10.1038/s41586-022-05431-4

[32] Ngo TD, Huynh T, Jung H, Ali F, Jeon J, Choi MS, et al. Modulation of contact resistance of dual-gated MoS_2 FETs using Fermi-level pinning-free antimony semi-metal contacts. Advancement of Science. 2023;**10**(21):2301400. DOI: 10.1002/advs.202301400

[33] Hoang L, Khan AI, Bennett RKA, Kim H-M, Zhang Z, Hocking M, et al. Enabling P-Type Conduction in Bilayer WS_2 with NbP Topological Semimetal Contacts. 2024. DOI: 10.48550/arXiv.2409.18926. Available from: https://arxiv.org/abs/2409.18926

[34] Pető J, Ollár T, Vancsó P, Popov ZI, Magda GZ, Dobrik G, et al. Spontaneous doping of the basal plane of MoS_2 single layers through oxygen substitution under ambient conditions. Nature Chemistry. 2018;**10**(12):1246-1251. DOI: 10.1038/s41557-018-0136-2

[35] Ngo TD, Choi MS, Lee M, Ali F, Hassan Y, Ali N, et al. Selective electron beam patterning of oxygen-doped WSe_2 for seamless lateral junction transistors. Advancement of Science. 2022;**9**(26):2202465. DOI: 10.1002/advs.202202465

[36] Jeong T, Kim J, Kim UJ, Ji H, Yun SJ. Oxidation effects on the optical and electrical properties of MoS_2 under controlled baking temperatures. Current Applied Physics. 2025;**70**:61-68. DOI: 10.1016/j.cap.2024.11.013

[37] Huynh T, Ngo TD, Choi H, Choi M, Lee W, Nguyen TD, et al. Analysis of p-type doping in graphene induced by monolayer-oxidized TMDs. ACS Applied Materials & Interfaces. 2024;**16**(3):3694-3702. DOI: 10.1021/acsami.3c16229

[38] Lee K, Kim BSY, Choi MS, Ali N, Shin H, Yue D, et al. Double-side-doped 2D TMD FET channel. ACS Applied Electronic Materials. 2024;**6**(5):3894-3900. DOI: 10.1021/acsaelm.4c00506

[39] Choi MS, Nipane A, Kim BSY, Ziffer ME, Datta I, Borah A, et al. High carrier mobility in graphene doped using a monolayer of tungsten oxyselenide. Nature Electronics. 2021;**4**(10):731-739. DOI: 10.1038/s41928-021-00657-y

[40] Wang P, Gao B, Liu W. *In situ* doping effect in monolayer MoS_2 via laser irradiation. Nano Ex. 2024;**5**(1):015018. DOI: 10.1088/2632-959X/ad2b84

[41] Tessarek C, Gridenco O, Wiesing M, Müssener J, Figge S, Sebald K, et al. Controlled laser-thinning of MoS_2 nanolayers and transformation to amorphous MoO_x for 2D monolayer fabrication. ACS Applied Nano Materials. 2020;**3**(8):7490-7498. DOI: 10.1021/acsanm.0c01104

[42] Kim BSY, Ngo TD, Hassan Y, Chae SH, Yoon SG, Choi MS. Advances

and applications of oxidized van der Waals transition metal dichalcogenides. Advancement of Science. 2024;**11**(43): 2407175. DOI: 10.1002/advs.202407175

[43] Huo J, Xiao Y, Sun T, Zou G, Shen D, Feng B, et al. Femtosecond laser irradiation-mediated MoS_2-metal contact engineering for high-performance field-effect transistors and photodetectors. ACS Applied Materials & Interfaces. 2021;**13**(45):54246-54257. DOI: 10.1021/acsami.1c12685

[44] Jang J, Kim JK, Shin J, Kim J, Baek KY, Park J, et al. Reduced dopant-induced scattering in remote charge-transfer-doped MoS_2 field-effect transistors. Science Advances. 2022;**8**(38):eabn3181. DOI: 10.1126/sciadv.abn3181

[45] Jeong I, Cho K, Yun S, Shin J, Kim J, Kim GT, et al. Tailoring the electrical characteristics of MoS_2 FETs through controllable surface charge transfer doping using selective inkjet printing. ACS Nano. 2022;**16**(4):6215-6223. DOI: 10.1021/acsnano.2c00021

[46] McClellan CJ, Yalon E, Smithe KKH, Suryavanshi SV, Pop E. High current density in monolayer MoS_2 doped by AlO_x. ACS Nano. 2021;**15**(1):1587-1596. DOI: 10.1021/acsnano.0c09078

[47] Jiang J, Xu L, Du L, Li L, Zhang G, Qiu C, et al. Yttrium-doping-induced metallization of molybdenum disulfide for ohmic contacts in two-dimensional transistors. Nature Electronics. 2024;**7**(7):545-556. DOI: 10.1038/s41928-024-01176-2

[48] Zhang S, Hill HM, Moudgil K, Richter CA, Hight Walker AR, Barlow S, et al. Controllable, wide-ranging n-doping and p-doping of monolayer group 6 transition-metal disulfides and diselenides. Advanced Materials. 2018;**30**(36):1802991. DOI: 10.1002/adma.201802991

[49] Ngo TD, Huynh T, Moon I, Taniguchi T, Watanabe K, Choi MS, et al. Self-aligned top-gate structure in high-performance 2D p-FETs via van der Waals integration and contact spacer doping. Nano Letters. 2023;**23**(23):11345-11352. DOI: 10.1021/acs.nanolett.3c04009

[50] Xue J-Y, Li F-L, Zhao Z-Y, Ni C-Y, Gu H-W, Young DJ, et al. In situ generation of bifunctional Fe-doped MoS_2 nanocanopies for efficient electrocatalytic water splitting. Inorganic Chemistry. 2019;**58**(16):11202-11209. DOI: 10.1021/acs.inorgchem.9b01814

[51] Sovizi S, Szoszkiewicz. Single atom doping in 2D layered MoS_2 from a periodic table perspective. Surface Science Reports. 2022;**77**(3):100567. DOI: 10.1016/j.surfrep.2022.100567

[52] Yang L, Cui X, Zhang J, Wang K, Shen M, Zeng S, et al. Lattice strain effects on the optical properties of MoS_2 nanosheets. Scientific Reports. 2014;**4**:5649. DOI: 10.1038/srep05649

[53] Qin S, Lei W, Liu D, Chen Y. In-situ and tunable nitrogen-doping of MoS_2 nanosheets. Scientific Reports. 2014;**4**:7582. DOI: 10.1038/srep07582

[54] Manzeli S, Ovchinnikov D, Pasquier D, Yazyev OV, Kis A. 2D transition metal dichalcogenides. Nature Reviews Materials. 2017;**2**:17033. DOI: 10.1038/natrevmats.2017.33

[55] Ghosh S, Zheng Y, Zhang Z, Sun Y, Schranghamer TF, Sakib NU, et al. Monolithic and heterogeneous three-dimensional integration of two-dimensional materials with high-density vias. Nature Electronics. 2024;**7**:892-903. DOI: 10.1038/s41928-024-01251-8

[56] Tang J, Wei Z, Wang Q, Wang Y, Han B, Li X, et al. In situ oxygen doping of monolayer MoS_2 for novel electronics. Small. 2020;**16**(42):2004276. DOI: 10.1002/smll.202004276

[57] Fu S, Kang K, Shayan K, Yoshimura A, Dadras S, Wang X, et al. Enabling room temperature ferromagnetism in monolayer MoS_2 via in situ iron-doping. Nature Communications. 2020;**11**:2034. DOI: 10.1038/s41467-020-15877-7

Chapter 4

P-Channel MoS_2 Transistors: Enabling Complementary Logic Circuits for Next-Generation Electronics

Zichao Ma and Changjian Zhou

Abstract

A comprehensive overview of recent progress in p-channel MoS_2 transistors and their integration into complementary logic circuits was provided in this chapter. The key challenges in developing high-performance p-type devices were highlighted, including reliable p-type doping techniques, efficient current polarity control, and minimizing contact resistance. Various doping strategies were systematically analyzed, focusing on their impact on unipolar conduction, on/off current ratios, and carrier mobility. A performance benchmark was presented to outline a clear path toward scalable, high-performance p-type MoS_2 transistors. The discussion emphasized the co-optimization of device physics, structural design, and fabrication methods, with particular attention to contact engineering and dielectric interface integration. Furthermore, recent implementations of complementary logic circuits highlighting novel device architectures and three-dimensional integration approaches were reviewed. These advancements underscored the promise of p-type MoS_2 transistors for enabling energy-efficient, high-density logic circuits based on two-dimensional materials.

Keywords: MoS_2, p-type, doping, field-effect transistor, complementary circuit

1. Introduction

The discovery of two-dimensional (2D) materials has transformed the landscape of nanoelectronics. Among these materials, molybdenum disulfide (MoS_2) stands out as a promising candidate for advanced field-effect transistor (FET) technologies, owing to its atomic-scale thickness, sizable bandgap, and excellent carrier mobility [1, 2]. MoS_2 has been widely investigated for FETs, where its ultrathin channel enables effective gate modulation of charge carriers, offering a clear path toward ultra-miniaturized devices [3–5]. Despite significant progress in n-channel MoS_2 FETs, the development of p-channel MoS_2 FETs has lagged behind, primarily due to the lack of robust and effective engineering techniques to reliably tailor the electronic properties of MoS_2 active layers for circuit integration and optoelectronic applications [6–9].

P-channel devices are indispensable for integrating functional circuits that rely on complementary configurations. In recent years, significant efforts have been dedicated to establishing a roadmap for engineering MoS_2 into high-performance p-channel devices and complementary circuits. As illustrated in **Figure 1**, this development involved several key stages: p-type MoS_2 synthesis, fabrication of p-type MoS_2 transistors, and performance optimization, ultimately enabling complementary integration. Doping MoS_2 to achieve p-type conduction became the critical first step for controlling carrier polarity and conductivity [10, 11]. Through the incorporation of dopant species or hybridization with other materials, researchers have demonstrated notable improvements in device characteristics and the realization of novel diode and transistor structures [12–14]. Careful modulation of the doping process allowed for precise control over polarity and conductivity, thereby tailoring material properties to meet the demands of various electronic and optoelectronic applications [15–17].

This chapter has been organized to present, in subsequent sections, practical methods for synthesizing p-type MoS_2, including substitutional doping *via* chemical vapor deposition (CVD) and charge transfer-based techniques that are compatible with the standard complementary metal-oxide-semiconductor (CMOS) fabrication process. These methods offered precise control over the lattice structure and electronic properties, enabling the customization of MoS_2 for complementary logic applications. Following this, we examined the characteristics of high-performance p-channel MoS_2 FETs, particularly emphasizing the critical role of contact and interface engineering. Finally, the chapter discussed the integration of these devices into complementary circuits, highlighting their potential for energy-efficient and high-density integrated logic systems.

2. Synthesis of P-type MoS_2 films

MoS_2 with atomic thickness is highly sensitive to environmental influences, making its electronic properties easily tunable through external factors. Early studies revealed significant doping effects in 2D MoS_2, prompting extensive exploration of doping strategies. Gate dielectrics and surface treatment using ions or molecules have been shown to modulate conductivity *via* charge transfer, although such approaches

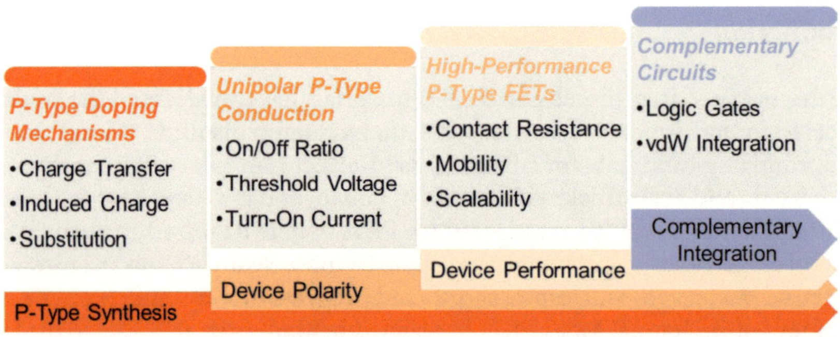

Figure 1.
Roadmap of MoS_2 for p-type FETs and complementary circuits.

may face stability and compatibility issues during circuit integration [18–20]. In contrast, substitutional doping, analogous to silicon (Si) technology techniques, offered more stable doping and was promising for scalable, wafer-level fabrication [21–23]. However, the limitations of traditional ion implantation for atomically thin 2D materials have led to alternative surface modification methods and *in situ* doping during synthesis [24–26]. The established synthesis techniques for producing large-area, high-quality p-type MoS_2 films were reviewed in this section.

2.1 Charge transfer-based doping methods

Charge transfer-induced p-type doping occurs when external acceptor states withdraw electrons from MoS_2, shifting its Fermi level toward the valence band. This phenomenon is intrinsic to 2D layered materials. Electron exchange can be triggered by adatoms [27], dielectric dangling bonds [28], and absorbed molecules [29]. As a result, various chemical methods, such as solution soaking, dielectric deposition, and plasma treatment, have been developed to exploit and control this charge transfer process effectively [19, 25, 30].

$AuCl_3$ chemical doping enabled the conversion of inherently n-type multilayer MoS_2 to p-type through controlled charge transfer [31, 32]. As shown in **Figure 2(a)**, this method employed mechanically exfoliated MoS_2 flakes with a thickness of about 10 nm, and then spin-coated with a 5 mM $AuCl_3$ solution at room temperature, followed by annealing at 100°C for 10 minutes in the air to activate the dopant. The resulting device structure of the p-channel MoS_2 FET is shown in **Figure 2(b)**. $AuCl_3$ acts as a strong electron acceptor, extracting electrons and inducing hole accumulation within the top 1.5–3 nm of the MoS_2 film, effectively shifting the Fermi level toward the valence band without disrupting the van der Waals structure. The doping outcomes depend on parameters such as $AuCl_3$ concentration, MoS_2 thickness, and annealing conditions. Selective doping was achievable by masking designated regions

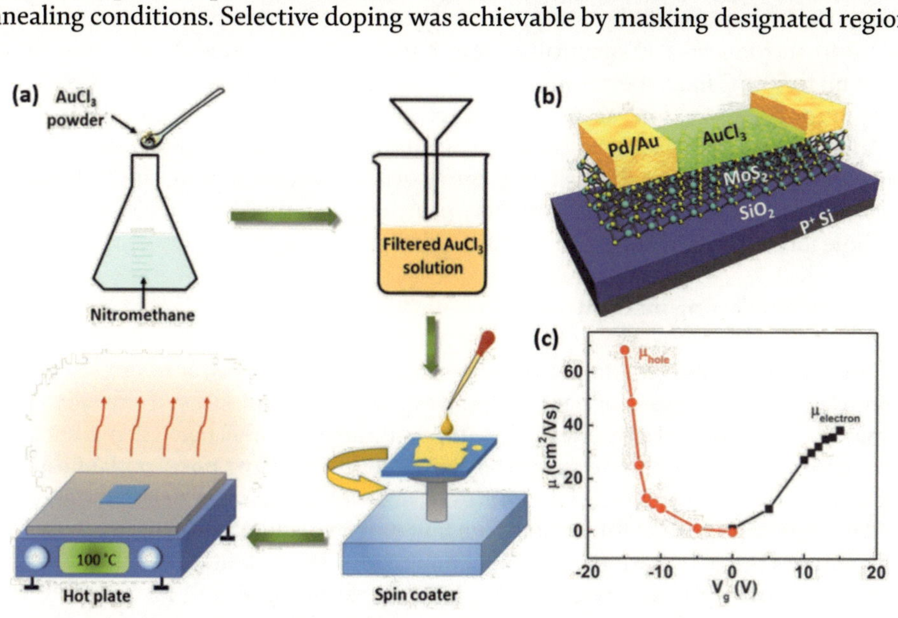

Figure 2.
(a) P-type doping of MoS_2 using an $AuCl_3$ treatment (Reproduced with permission from [32], Springer Nature, 2022) (b) Device structure of $AuCl_3$ solution doped p-channel MoS_2 FET; (c) Mobility of p-channel and n-channel MoS_2 FETs (Reproduced with permission from [31], John Wiley and Sons, 2016).

during spin coating, and the resulting p-type behavior remained stable for over a month under ambient conditions. The fabricated FETs exhibited a clear transition from n-type to p-type conduction and hole mobility exceeding 60 $cm^2/(V{\cdot}s)$. Such high hole mobility was identified as one key advantage of charge transfer doping [33].

While $AuCl_3$ demonstrated effectiveness for multilayer MoS_2, its impact on monolayer MoS_2 is limited. To address this, tris(4-bromophenyl)-ammoniumyl hexachloroantimonate (Magic Blue), a molecular dopant with stronger charge transfer capabilities, has been used to achieve p-type doping in monolayer MoS_2 [34]. With a high electron affinity (~5.5 eV) closely matching the valence band of monolayer MoS_2, Magic Blue efficiently extracted electrons and enabled hole accumulation in the monolayer MoS_2. The doping process was performed by immersing the monolayer MoS_2 FETs in a Magic Blue solution in dichloromethane (DCM) for 30 ~ 60 seconds, resulting in a uniform p-type doping stable in air for several days. Following doping, devices exhibited a positive threshold voltage shift, enhanced hole injection, and suppressed n-type conduction. The Schottky barrier for holes was also lowered, reducing contact resistance at the metal–MoS_2 interface. As a result, Magic Blue not only acts as a dopant but also modifies contacts, contributing to the overall improved performance of p-type MoS_2 FETs. Notably, the doping effect was partially reversible. A vacuum annealing process at around 200°C has removed Magic Blue, allowing tunable doping and potential device reusability.

In addition, plasma-assisted doping techniques have emerged as a scalable and robust alternative [24]. In particular, oxygen plasma treatment under a toroidal magnetic field has proven to induce stable p-type conduction in MoS_2 with minimal lattice damage and improved mobility [25]. Moreover, this process was compatible with the standard CMOS fabrication process, enabling practical process integration. The toroidal magnetic field effectively reduced ion energy and prolonged the residence time of reactive oxygen species on the surface of exposed MoS_2 films, allowing for controlled doping. During treatment, both negative oxygen ions (O^-) and positive oxygen ions (O_2^+) were introduced and substituted sulfur atoms at defect sites. A post-annealing step at up to 500°C in an inert atmosphere healed defects and stabilized oxygen incorporation. The O_2^+ ions act as shallow acceptors, shifting the Fermi level toward the valence band. The magnetic field enhanced the O_2^+/O^- ratio, suppressing MoO_x formation and preserving the lattice integrity of MoS_2. After doping, the MoS_2 work function increased from 4.3 to 5.1 eV. The resulting p-channel MoS_2 FETs exhibited high hole mobility, reaching up to 115 $cm^2/(V{\cdot}s)$.

2.2 Controlled doping method

Conventional CMOS ion implantation techniques, which produced doping profiles extending tens to hundreds of nanometers deep, were less suited for atomically thin 2D materials like MoS_2. High-energy ion implantation not only lacked depth control in 2D materials but also caused severe lattice damage, requiring post-implantation annealing at temperatures exceeding 400°C. Such conditions were detrimental to the van der Waals structure of MoS_2, often leading to irreversible degradation.

To address these limitations, a plasma immersion ion implantation (PIII) method has been developed for selective, low-energy phosphorus doping of MoS_2 [24]. Using an inductively coupled plasma (ICP) source with a PH_3/He gas mixture, phosphorus ions were generated and implanted under a pulsed DC bias (0–2 keV), which confined doping to the top 5 nm of the MoS_2 film. The pulsed bias minimized surface damage and allowed for area-selective doping. A post-implantation annealing at

300°C helped activate dopants and repair implantation-induced defects. Phosphorus doping occurred primarily *via* substitution at sulfur vacancies, leading to intralayer charge transfer that shifted the Fermi level toward the valence band. The doping level was tuned by adjusting implantation energy and dose, achieving sheet hole concentrations from 10^{10} to 10^{12} cm^{-2} and hole mobilities ranging from 8.4 to 137.7 $cm^2/(V{\cdot}s)$, depending on doping degeneracy.

Substitution of sulfur atoms by oxygen in MoS_2 has been extensively explored through both high-temperature and ambient oxidation methods [27, 35–37]. Under ambient conditions (air, room temperature, and light), oxygen atoms can spontaneously incorporate into the basal plane of monolayer MoS_2, forming a highly crystalline molybdenum oxy-sulfide (MoS_xO_y) phase while maintaining the original lattice structure. This slow, substitutional process gradually resulted in solid-solution-type MoS_xO_y crystals, where sulfur was progressively replaced by oxygen. Atomic-scale characterization *via* scanning tunneling microscopy (STM) revealed dark triangular features corresponding to sulfur vacancies and bright spots indicating oxygen substitution. These observations were supported by density functional theory (DFT) simulations, confirming the substitutional nature of the doping. Notably, the oxidation process was reversible. Annealing in H_2S at 200°C can fully restore pristine MoS_2. While the process was thermodynamically favorable for MoS_2 (activation energy ΔE (=) −0.49 eV), it was not for $MoSe_2$ (ΔE = +0.75 eV). Additionally, the kinetic barrier of ~1 eV allowed the oxidation of MoS_2 to proceed slowly over the course of months under ambient conditions.

Dopants from elements to the left of Mo in the periodic table, such as niobium (Nb), vanadium (V), and tantalum (Ta), have been shown to effectively induce p-type doping in MoS_2 across both monolayer and multilayer forms [38–41]. Among these, Nb has been identified as a particularly effective acceptor dopant. For example, a layered Mo (2.5 nm)/Nb/Mo (2.5 nm) structure deposited *via* e-beam evaporation has been converted into Nb-doped MoS_2 through sulfurization at 900°C [42]. Varying the Nb layer thickness (e.g., 0–0.3 nm) resulted in degenerate doping levels with Nb concentrations up to 1.5×10^{21} cm^{-3}. However, heavy doping came at the cost of reduced mobility, with values around 0.5 $cm^2/(V{\cdot}s)$. In contrast, moderate doping achieved higher mobilities up to 8.5 $cm^2/(V{\cdot}s)$. Notably, Nb-doped MoS_2 film exhibited surface roughness of 1.3 nm and low contact resistance (0.6 kΩ·μm with Ni contacts) while retaining crystallinity, as confimed by X-ray diffraction (XRD) characterizations.

As shown in **Figure 3(a)** and **(b)**, structural analysis using transmission electron microscope (TEM) and convergent beam electron diffraction measurement revealed that Nb substitution has transformed the stacking order of MoS_2 from the natural 2H phase to the 3R phase [14]. DFT simulations confirmed this transition as thermodynamically favorable. The 3R phase lacked inversion symmetry, making it promising for valleytronic applications. Optical measurements indicated that the bandgap of monolayer MoS_2 remained near 1.8 eV. In contrast, Nb doping in multilayer MoS_2 hybridized with valence band density of states near the Γ point, reducing the indirect bandgap through a band-anticrossing mechanism. In addition, Nb substitution in monolayer MoS_2 formed localized in-gap states, leading to strong broadband photoluminescence (PL) from exciton complexes. The PL exhibited increased intensity and extended lifetimes compared to undoped MoS_2, along with slight redshifts in exciton transitions due to the structural phase change.

A one-step salt-assisted CVD process has been developed for the wafer-scale synthesis of uniform Nb-doped monolayer MoS_2 films [43, 44]. As shown in **Figure 3(c)**, a precursor mixture containing Nb_2O_5 (dopant precursor) and NaCl

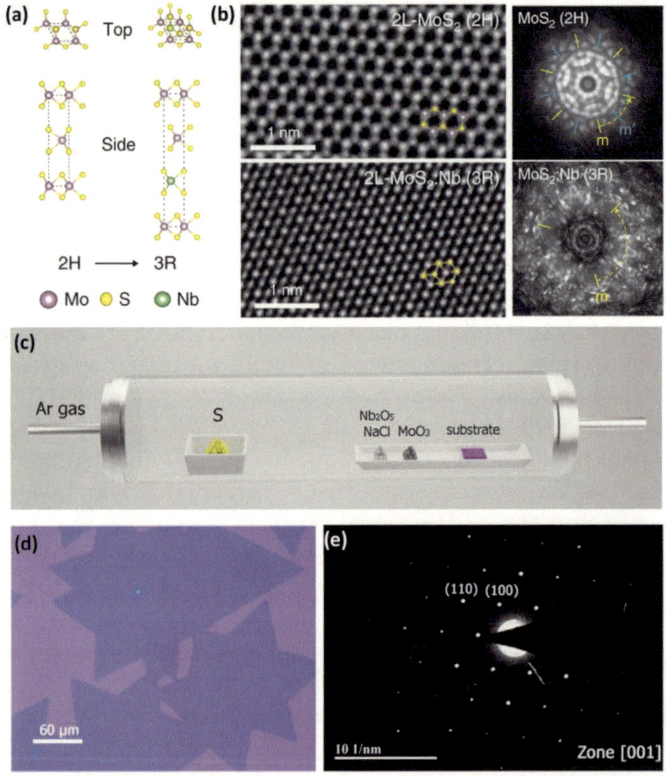

Figure 3.
(a) Atomic structure of undoped 2H-MoS_2 and Nb-doped p-type 3R-MoS_2 lattices. (b) TEM images and convergent beam electron diffraction patterns of undoped MoS_2 and p-type-doped-MoS_2 lattices (Reproduced with permission from [14], Springer Nature, 2018). (c) Chemical vapor deposition method to synthesize large-area monolayer MoS_2 films. (d) Optical image of CVD-synthesized Nb-doped p-type MoS_2 monolayer films. (e) TEM selected area electron diffraction image of the CVD-synthesized Nb-doped p-type MoS_2 monolayer (Reproduced with permission from [43], American Chemical Society, 2020).

(used as a growth promoter) was combined with MoO_3 and heated in a high-temperature furnace at 800°C, while sulfur vapor was introduced by an Ar gas flow at 210°C. The synthesized monolayer films were characterized by optical measurements, TEM, and selected area electron diffraction (SAED), as shown in **Figure 3(d)** and **(e)**, confirming successful doping and structural integrity. Furthermore, DFT simulations have confirmed that Nb substitution introduced acceptor states near the valence band maximum, shifted the Fermi level downward, and suppressed the intrinsic n-type conduction of MoS_2. Compared to methods such as chemical vapor transport (CVT) or mechanical exfoliation, the CVD approach offered simplicity, scalability, and the ability to produce large-area, high-quality doped monolayer MoS_2 films. This tunability was considered critical for enabling complementary electronic applications.

3. High-performance P-type MoS_2 FETs and circuits

Compared to n-type devices, p-channel MoS_2 FETs often suffer from inefficient hole injection and limited hole transport [45–47]. Overcoming these limitations requires comprehensive optimization in contact engineering [48–51], mobility enhancement [52–54], and dielectric interface quality [55–57]. This section reviews recent advancements in p-type MoS_2 FETs, focusing on key device metrics, such as contact resistance, mobility, and on/off current ratio, which are critical for energy-efficient complementary device integration. This section highlights emerging efforts to integrate p-channel MoS_2 devices with n-channel counterparts in van der Waals (vdW) structures, enabling complex logic gate circuit implementation.

3.1 P-channel MoS_2 FETs

The electrical performance of p-type MoS_2 FETs strongly depended on the effectiveness of the doping strategy used to induce hole conduction. Using molecular adsorbates or elemental dopants, charge transfer-based doping has emerged as a non-invasive and effective method for enabling p-type behavior [31, 58]. These techniques offered advantages such as compatibility with monolayer MoS_2, high doping concentrations, and, in many cases, hole mobilities. However, the performance of charge transfer doping was susceptible to process parameters [25, 36, 59]. Moreover, small fluctuations in dopant concentration caused unintended polarity shifts. These sensitivities highlighted broader challenges in achieving stable, uniform, and scalable doping, which were key requirements for integrating large-area, reproducible, large-scale integrated circuits.

Substitutional doping, where Mo atoms were replaced with elements such as Nb, Ta, or V, offers fixed acceptor levels near the valence band, enabling controlled and stable p-type behavior [38–41]. Nb-doped MoS_2 was among the first to be synthesized, allowing exfoliation of single-crystalline p-type flakes across monolayer to bulk thicknesses [60]. This has supported the systematic exploration of device physics and optimization strategies. Several studies have since employed exfoliated Nb-doped MoS_2 to fabricate unipolar p-channel FETs with high on/off current ratios.

FETs fabricated using exfoliated Nb-doped MoS_2 flakes (1.4–10 nm thick) with nickel (Ni) as the contact metal exhibited distinct electrical characteristics compared to undoped MoS_2 [61]. Notably, as the channel thickness increased, the device behavior shifted from accumulation mode to depletion mode. The on/off current ratio also decreased with thickness. For example, a 3 nm channel achieved an on/off ratio exceeding 10^4, whereas a 10 nm channel showed a reduced ratio of 10. This trend was attributed to the weakened electrostatic gate control in thicker channels. The Nb doping concentration in these flakes was estimated to be 3×10^{19} cm^{-3}, indicative of degenerate doping levels. While thinning the channel may improve switching behavior, it reduces the on-state current slightly. Further reduction to bilayer MoS_2 has resulted in ambipolar transport behavior.

A key challenge in achieving efficient hole injection in MoS_2 FETs was caused by the strong Fermi level pinning near the conduction band edge, even when using high work function metals [62–64]. This pinning was largely caused by metal-induced gap states (MIGS) and interfacial defect states at the MoS_2–metal interface, which has limited the tunability of the Schottky barrier height (SBH) through metal selection [65–67]. Without proper doping, the hole injection barrier (Φ_h) remained high, resulting in poor hole injection and elevated contact resistance (R_C). Consequently, undoped MoS_2 devices typically exhibited unipolar n-type conduction [68–70]. In contrast, doping introduced a significant density of states near the valence band edge, which could complement with the MIGS and defect states, allowing better modulation of the Fermi level at the metal–semiconductor interface [71–74].

As compared in **Figure 4 (a–c)**, studies that explored different contact metals have further clarified the role of Fermi level alignment [75, 77, 78]. When platinum (Pt) contacts were employed with Nb-doped MoS_2 flakes (3–8 nm thick) and a 30 nm Al_2O_3 gate dielectric (in contrast to SiO_2), devices exhibited similar transfer characteristics [61]. The extracted contact barrier height was around 0.1 eV in multilayer devices, and contact resistance was significantly lower than that with Ni contacts. Preliminary studies using additional metals such as Au and Ti have also shown suppressed electron conduction and enhanced hole transport in Nb-doped MoS_2 FETs. By contrast, Pt contacts on undoped MoS_2 typically formed small electron barriers (~0.2 eV) and supported unipolar n-type conduction with on/off ratios larger than 10^6. These findings suggested that Nb doping has shifted the Fermi level downward at the metal–MoS_2 interface, reducing Φ_h and facilitating hole transport. Nonetheless, doping alone has not demonstrated Ohmic contact.

Furthermore, the contact fabrication process also played a critical role in determining device performance. Techniques such as slow metal deposition and interfacial defect passivation have been shown to mitigate Fermi level pinning, enabling high-work function metals like Pt and Pd to align more closely with the MoS_2 valence band [45, 76, 79]. This alignment helped reduce the Schottky barrier height and contact resistance. Additionally, the choice of dopant species significantly impacted the minimum achievable contact resistance. For example, in degenerate p-type doping of $MoSe_2$, a 2D material similar to MoS_2, Nb-doped films exhibited contact resistances 2 ~ 8 times smaller than those of V-doped and Ta-doped films [22]. While first-principles simulations predicted only minor differences among the electronic structures of doped MoS_2, the observed variations in device performance suggested that dopant ionization and lattice structure may differ from theoretical expectations.

Further experimental studies were required to understand these effects fully.

Few-layer MoS_2 was particularly sensitive to its surrounding environment, and the application of external biasing through a thin and high-dielectric-constant (high-κ) gate dielectric could generate a significant hole carrier concentration [20, 57]. Such electrostatic doping *via* a polarity control gate structure enabled dynamic modulation

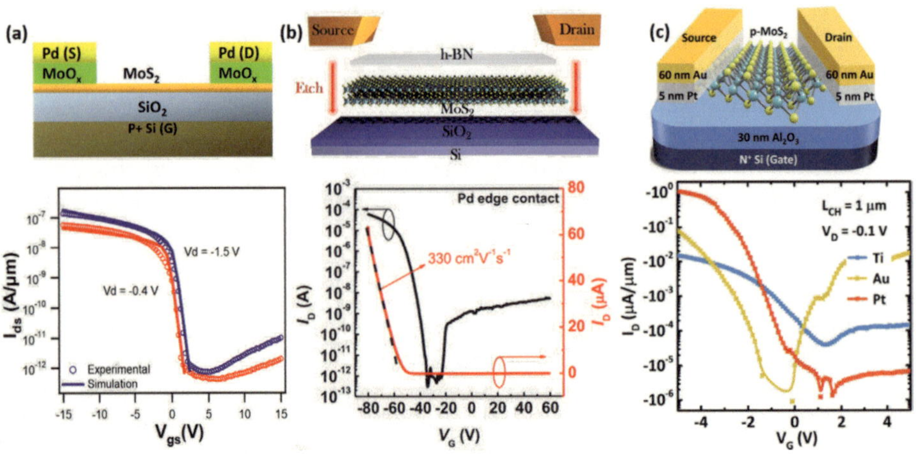

Figure 4.
Device structure and transfer characteristics of p-type MoS_2 FETs fabricated using (a) high work function MoO_x contact (Reproduced with permission from [75], American Chemical Society, 2014) (b) edge contact using Pd (Reproduced with permission from [76], John Wiley and Sons, 2019) and (c) top-contact using evaporated Pt, Au, and Ti electrodes and Nb-doped MoS_2 channels (Reproduced with permission from [61], IEEE, 2021).

of carrier type and concentration without permanently modifying the MoS_2 lattice. This technique facilitated direct observation of transitions from n-type to ambipolar and ultimately to unipolar p-type transport. Achieving such polarity reversal typically requires an intrinsically ambipolar material in conjunction with a control gate that substantially overlaps the contact region. This has been demonstrated in MoS_2 using a double-gate structure, where the top gate incorporated a bilayer SiO_2/HfO_2 dielectric stack. The SiO_2 layer, deposited using a plasma-enhanced process on a 5 nm-thick MoS_2 channel, introduced significant p-type charge transfer. Using the back gate and HfO_2 dielectric, the device exhibited n-type conduction, whereas the top gate with the SiO_2/HfO_2 stack reversed the behavior to p-type conduction, clearly demonstrating electrostatically induced polarity switching. The theoretical basis for the gate dielectric effect on MoS_2 FET performance has not yet been fully established.

Comprehensive studies examining the energy spectra of interface state densities (D_{it}) in both n-type and p-type MoS_2 across various thicknesses and dielectric stacks have revealed notable trends [80]. In undoped n-type MoS_2, the D_{it} values near mid-gap decreased from 5×10^{12} cm^{-2} eV^{-1} in high-κ top gate devices to 5×10^{11} cm^{-2} eV^{-1} in FETs utilizing h-BN dielectrics. In contrast, Nb-doped p-type MoS_2 consistently shows high D_{it} values even when integrated into h-BN dielectrics. These findings suggested that the interface effect originated from two primary sources: externally induced strain, which was caused by surface roughness and high-κ deposition processes, affected the conduction band, while sulfur-vacancy-related defect states dominated the valence band. Though mid-gap sulfur vacancy (V_S) states near the conduction band edge were highly localized, shallow V_S states near the valence band could hybridize with valence band states, especially in the case of disulfur vacancies or vacancy clusters [81–83]. Although Nb doping introduced shallow acceptor-like states, the significantly higher density of sulfur vacancies in atomically thin MoS_2 suggested that V_S defects were the dominant factor limiting hole transport. Therefore, enhancing the crystalline quality of 2D MoS_2 films became essential for improving hole conduction, particularly in the subthreshold regime.

Table 1 summarized key performance metrics of p-type MoS_2 FETs achieved using various doping techniques. In general, increased the doping concentration has led to higher hole current. However, this often came at the cost of reduced mobility and degraded subthreshold swing, primarily due to enhanced carrier scattering and deteriorated electrostatics.

3.2 Complementary MoS_2 circuits

Complementary logic, as the cornerstone of modern low-power digital electronics, relies on the co-integration of both p-type and n-type transistors [88, 89]. The integration of p-type MoS_2 FETs into complementary logic circuits has become an increasingly attainable goal. The monolithic integration of high-performance n-type and p-type transistors within the same layered MoS_2 platform was particularly desirable for next-generation electronic systems. As shown in **Figure 5(a)**, one promising approach involved the controlled growth of MoS_2 on fin-shaped oxide structures, coupled with heavily doped Si used as source and drain contacts. In this configuration, the fabrication of complementary MoS_2 FETs has been demonstrated using a conventional Si-compatible ion implantation process [84]. While the as-grown MoS_2 films remain undoped, the selective implantation effectively modulated the Fermi levels of the Si contact regions, thereby alleviating Fermi level pinning, significantly improving contact

Doping Concentration (cm^{-3}) and Method	Thickness (nm)	$I_P/I_{Off}/I_N$ (μA @ V_{DS} = 1 V)	R_C (kΩ•μm)	Mobility ($cm^2 \cdot V^{-1} \cdot s^{-1}$)	Ref.
2 × 10^{17} (P implantation)	3.5	$1/10^{-2}/10^{-1}$	/	137.7	[24]
/ (O_2^+ substitution)	6-layers	$1/10^{-7}/10^{-5}$	/	115.2	[25]
/ ($AuCl_3$ solution)	10	$10/10^{-7}/10^{-6}$	2.3	68	[31]
/ (Magic blue solution)	1-layer	$10^{-1}/10^{-6}/10^{-5}$	482 ± 40	2.3 ± 0.7	[34]
9.5 at% (V substitution)	1-layer	$1/10^{-5}/10^{-5}$	/	/	[40]
67% Ta: MoS_2 (Ta substitution)	1-layer	0.064/0.054/0.054	/	/	[41]
3 × 10^{19} (Nb substitution)	2.7	$1/10^{-4}/10^{-4}$	/	6.7	[60]
/ (High work function MoO_x contact)	40	$10^{-1}/10^{-6}/10^{-5}$	/	86	[75]
/ (vdW Pt contact)	/	$10/10^{-6}/10^{-1}$	/	175	[77]
/ (edge Pd contact)	7.5	$10^2/10^{-6}/10^{-2}$	/	330	[76]
/ (edge Au contact)	80	$10^1/10^{-5}/10^{-1}$	/	432	[76]
/ (Heavily doped Si contacts)	5	$1/10^{-6}/10^{-6}$	/	10	[84]
3 × 10^{19} (Nb substitution)	5	$10^1/10^{-5}/10^{-5}$	7	60	[85]
/ (IGZO charge transfer)	123.7	$10^1/10^{-2}/10^{-2}$	/	24.1	[86]
/ (CrOCl charge transfer)	7	$1/10^{-5}/10^{-5}$	8.8	425	[87]

Table 1.
Benchmark the p-type MoS_2 FET performance metrics.

performance, and enabling robust complementary logic circuits. The fabricated complementary FETs have demonstrated excellent switching characteristics, with an on/off current ratio exceeding 10^6 and ultra-low leakage currents in the picoampere range. Furthermore, a complementary MoS_2 inverter has been fabricated, exhibiting a voltage gain greater than 20. These results offered compelling evidence that complementary logic operation using 2D materials within a unified system was not only feasible but also highly promising.

To streamline the integration of complementary logic, a homogeneous approach was developed wherein both n-channel and p-channel FETs were fabricated using the same material system, for example, MoS_2 as the active layer and Pt as the contact electrode [85]. The n-channel devices utilized undoped MoS_2, while the p-channel devices employed Nb-doped MoS_2 to modulate the Fermi level at the contact interface. In undoped MoS_2, the Pt Fermi level was pinned approximately 0.2 eV below the conduction band minimum, while in Nb-doped MoS_2, it shifted to about 0.2 eV above the valence band maximum. This Schottky barrier height modulation resulted in nearly symmetric on-state currents for both devices. However, a threshold voltage mismatch arose. The p-FET operated in

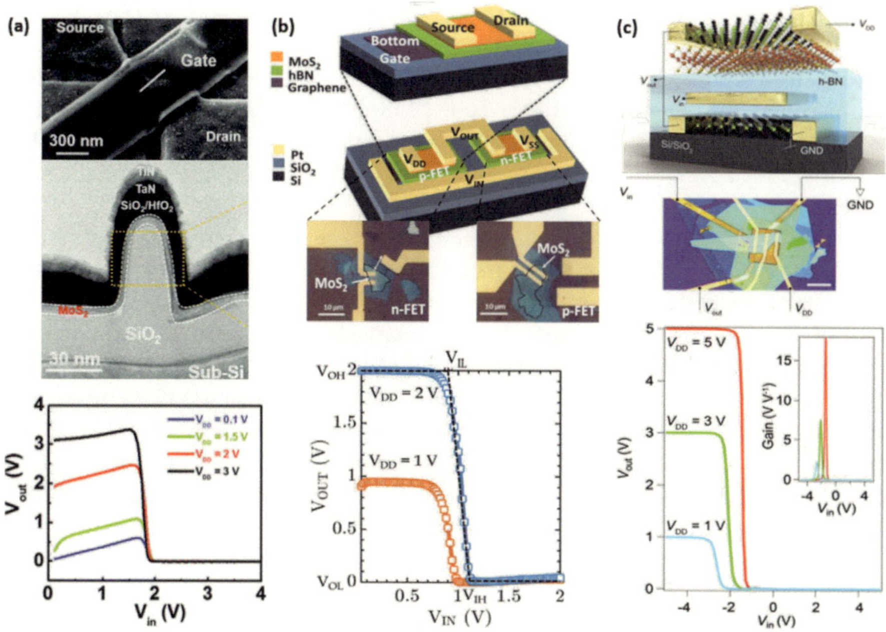

Figure 5.
FET device structure and voltage transfer characteristics of complementary MoS_2 inverters using (a) fin-shaped MoS_2 channel and heavily p-type and n-type doped Si contacts (Reproduced with permission from [84], Royal Society of Chemistry, 2019). (b) homogeneous integrated complementary MoS_2 FETs using all-2D gate capacitor stack and Pt contact electrodes (Reproduced with permission from [90], IEEE, 2021) and (c) 3D vdW stacked complementary FETs using n-type MoS_2 channel and CrOCl-doped p-type MoS_2 channels and Cr contacts (Reproduced with permission from [87], Springer Nature, 2024).

depletion mode with a threshold of about –2 V, while the n-FET operated in accumulation mode with a threshold near 0 V. Consequently, the inverter transition voltage was centered around –1 V, necessitating a dual power supply for standard logic gate operations.

To address this issue, an all-2D gate stack was introduced. As shown in **Figure 5(b)**, the FETs incorporate hexagonal boron nitride (h-BN) as the gate dielectric and graphene as the gate electrode in a bottom-gate configuration [90]. This 2D vdW structure effectively isolated the MoS_2 channel from substrate-induced doping effects, resulting in more stable and symmetric threshold voltages [91]. As a result, the n-FET and p-FET exhibited threshold voltages of approximately +2 V and –2 V, respectively, enabling complementary operation under one DC power supply. Both transistors demonstrated subthreshold swings of 0.21–0.23 V/dec, on/off current ratios exceeding 10^7, and output currents of 1 μA/μm at a channel length of 2 μm. The resulting complementary MoS_2 inverter achieved full rail-to-rail voltage transfer characteristics with a voltage gain 14 at a 2 V supply. Additionally, the symmetric characteristics led to ultra-low dynamic power consumption below 1 nW, nearly two orders of magnitude lower than the devices using Al_2O_3 dielectrics. This highlighted the energy efficiency and scalability of this homogeneous integration strategy.

In parallel, charge transfer-based doping methods have offered enhanced flexibility and compatibility with existing technologies. A notable example was a charge transfer technique that induced p-type conduction in MoS_2, which enabled the fabrication of complementary devices on a single flake [86]. In this method, an n-type indium gallium zinc oxide (IGZO) layer was deposited atop the MoS_2, followed by annealing in air. The doping level was tuned by adjusting the annealing time. The fabricated p-type MoS_2 FETs with Pt electrodes exhibited hole mobilities as high as 24.1 $cm^2/(V{\cdot}s)$. Leveraging this approach, a low-voltage complementary inverter was realized on a single MoS_2 flake, operating at a supply voltage of 1 V.

One particularly promising advancement toward realizing complementary 2D logic was the three-dimensional (3D) integration of n-channel and p-channel FETs into vertically stacked architectures. This vertical configuration enabled the implementation of more complex circuits through charge transfer doping within stacked van der Waals (vdW) heterostructures [87, 92]. As shown in **Figure 5(c)**, a notable example was the introduction of a vdW interfacial doping strategy that employed antiferromagnetic chromium oxychloride (CrOCl) to render inherently n-type 2D transition metal dichalcogenides (TMDs), including MoS_2, WSe_2, and $MoSe_2$, into stable, high-performance p-type FET devices. By stacking MoS_2 layers onto CrOCl, strong interfacial coupling facilitated charge transfer and induced favorable band alignment shifts, as confirmed by density functional theory calculations. This technique yielded p-type MoS_2 FETs with room temperature hole mobilities reaching up to 425 $cm^2/(V{\cdot}s)$ and on/off current ratios exceeding 10^6.

In pursuing 3D integration, fabricating ultrathin n-type and p-type 2D layers has been essential to maximize vertical device density. One study demonstrated a method to form vertically stacked junctions by chemically doping the top and bottom surfaces of mechanically exfoliated MoS_2 flakes using $AuCl_3$ to induce p-type behavior and benzyl viologen (BV) for n-type characteristics, which resulted in a controlled doping depth of approximately 1.5 nm [93]. The thickness of the MoS_2 film strongly influenced the electrical performance of these junctions. For example, flakes thicker than 8 nm exhibit conventional diode-like rectification under forward bias, while films thinner than 8 nm show tunneling-dominated "reversed" rectification under reverse bias. The practical limit for forming a stable junction was around 3 nm, corresponding to approximately four MoS_2 layers. Such vertical configuration supported the development of reconfigurable circuits and highlighted the potential of chemically doped MoS_2 p-n junctions in flexible optoelectronics.

4. Conclusions

Developing high-performance p-type MoS_2 transistors has been a key step toward achieving fully complementary 2D logic circuits. Advances in substitutional doping and charge transfer doping, contract engineering, and dielectric integration have enabled robust hole conduction with high mobility, large on/off ratios, and scalable fabrication. Recent demonstrations of integrated n-type and p-type MoS_2 FETs, with both in-plane and vertically stacked device structures, confirmed the feasibility of fully complementary logic within the MoS_2 material system. Techniques such as all-2D gate stacks, interfacial charge transfer doping, and chemically defined vertical junctions have further expanded the design space for low-power, high-density electronics. Despite challenges remaining in threshold control, large-area doping, and process integration, the progress outlined in this chapter highlights the strong potential of p-channel MoS_2 transistors as a building block for next-generation electronic systems.

Acknowledgements

This research was funded by the Guangdong Provincial Key Field Research and Development Program 2022B0701180002, the Guangdong Basic and Applied Basic Research Foundation 2025A1515010391, and the Guangzhou Science and Technology Program SL2024A04J3755.

Conflict of interest

The authors declare no conflict of interest.

Author details

Zichao Ma and Changjian Zhou*
School of Microelectronics, South China University of Technology, Guangzhou, China

*Address all correspondence to: zhoucj@scut.edu.cn

References

[1] Manzeli S, Ovchinnikov D, Pasquier D, Yazyev OV, Kis A. 2D transition metal dichalcogenides. Nature Reviews Materials. 2017;**2**:17033. DOI: 10.1038/natrevmats.2017.33

[2] Chhowalla M, Jena D, Zhang H. Two-dimensional semiconductors for transistors. Nature Reviews Materials. 2016;**1**:16052. DOI: 10.1038/natrevmats.2016.52

[3] Radisavljevic B, Radenovic A, Brivio J, Giacometti V, Kis A. Single-layer MoS_2 transistors. Nature Nanotechnology. 2011;**6**:147-150. DOI: 10.1038/nnano.2010.279

[4] Desai SB, Madhvapathy SR, Sachid AB, Llinas JP, Wang Q, Ahn GH, et al. MoS_2 transistors with 1-nanometer gate lengths. Science. 1979;**2016**(354):99-102. DOI: 10.1126/science.aah4698

[5] Fan D, Li W, Qiu H, Xu Y, Gao S, Liu L, et al. Two-dimensional semiconductor integrated circuits operating at gigahertz frequencies. Nature Electronics. 2023;**6**:879-887. DOI: 10.1038/s41928-023-01052-5

[6] Sheng C, Dong X, Zhu Y, Wang X, Chen X, Xia Y, et al. Two-dimensional semiconductors: From device processing to circuit integration. Advanced Functional Materials. 2023;**33**:2304778. DOI: 10.1002/adfm.202304778

[7] Dubey M, Li L-J, Chin ML, Kong J, Palacios T, Yu L, et al. Integrated circuits based on bilayer MoS_2 transistors. Nano Letters. 2012;**12**:4674-4680. DOI: 10.1021/nl302015v

[8] Yin L, Cheng R, Ding J, Jiang J, Hou Y, Feng X, et al. Two-dimensional semiconductors and transistors for future integrated circuits. ACS Nano. 2024;**18**:7739-7768. DOI: 10.1021/acsnano.3c10900

[9] Lu Y-C, Huang J-K, Chao K-Y, Li L-J, Hu VP-H. Projected performance of Si- and 2D-material-based SRAM circuits ranging from 16 nm to 1 nm technology nodes. Nature Nanotechnology. 2024;**19**:1066-1072. DOI: 10.1038/s41565-024-01693-3

[10] Loh L, Zhang Z, Bosman M, Eda G. Substitutional doping in 2D transition metal dichalcogenides. Nano Research. 2021;**14**:1668-1681. DOI: 10.1007/s12274-020-3013-4

[11] Tang L, Xu R, Tan J, Luo Y, Zou J, Zhang Z, et al. Modulating electronic structure of monolayer transition metal dichalcogenides by substitutional Nb-doping. Advanced Functional Materials. 2021;**31**:2006941. DOI: 10.1002/adfm.202006941

[12] Gao H, Suh J, Cao MC, Joe AY, Mujid F, Lee K-H, et al. Tuning electrical conductance of MoS_2 monolayers through substitutional doping. Nano Letters. 2020;**20**:4095-4101. DOI: 10.1021/acs.nanolett.9b05247

[13] Bhowmik G, Gruenewald K, Malladi G, Mowll T, Ventrice C, Huang M. Tunable photoluminescence of atomically thin MoS_2 via Nb doping. MRS Advances. 2019;**4**:609-614. DOI: 10.1557/adv.2019.24

[14] Suh J, Tan TL, Zhao W, Park J, Lin D-Y, Park T-E, et al. Reconfiguring crystal and electronic structures of MoS_2 by substitutional doping. Nature Communications. 2018;**9**:199. DOI: 10.1038/s41467-017-02631-9

[15] Li X, Yang J, Sun H, Huang L, Li H, Shi J. Controlled synthesis and accurate doping of wafer-scale 2D semiconducting transition metal dichalcogenides. Advanced Materials. 2023:2305115. DOI: 10.1002/adma.202305115

[16] Tarasov A, Zhang S, Tsai M-Y, Campbell PM, Graham S, Barlow S, et al. Controlled doping of large-area trilayer MoS_2 with molecular reductants and oxidants. Advanced Materials. 2015;**27**:1175-1181. DOI: 10.1002/adma.201404578

[17] Wang Z, Tripathi M, Golsanamlou Z, Kumari P, Lovarelli G, Mazziotti F, et al. Substitutional p-type doping in NbS_2–MoS_2 lateral heterostructures grown by MOCVD. Advanced Materials. 2023;**35**:2209371. DOI: 10.1002/adma.202209371

[18] Kiriya D, Tosun M, Zhao P, Kang JS, Javey A. Air-stable surface charge transfer doping of MoS_2 by benzyl viologen. Journal of the American Chemical Society. 2014;**136**:7853-7856. DOI: 10.1021/ja5033327

[19] Yang L, Majumdar K, Liu H, Du Y, Wu H, Hatzistergos M, et al. Chloride molecular doping technique on 2D materials: WS_2 and MoS_2. Nano Letters. 2014;**14**:6275-6280. DOI: 10.1021/nl502603d

[20] McClellan CJ, Yalon E, Smithe KKH, Suryavanshi SV, Pop E. High current density in monolayer MoS_2 doped by AlO_x. ACS Nano. 2021;**15**:1587-1596. DOI: 10.1021/acsnano.0c09078

[21] Li H, Yang J, Li X, Luo Q, Cheng M, Feng W, et al. Bridging synthesis and controllable doping of monolayer 4 in length transition-metal dichalcogenides single crystals with high electron mobility. Advanced Materials. 2023;**35**:2211536. DOI: 10.1002/adma.202211536

[22] Pan Y, Jian T, Gu P, Song Y, Wang Q, Han B, et al. Precise p-type and n-type doping of two-dimensional semiconductors for monolithic integrated circuits. Nature Communications. 2024;**15**:9631. DOI: 10.1038/s41467-024-54050-2

[23] Das M, Sen D, Sakib NU, Ravichandran H, Sun Y, Zhang Z, et al. High-performance p-type field-effect transistors using substitutional doping and thickness control of two-dimensional materials. Nature Electronics. 2024;**8**:24-35. DOI: 10.1038/s41928-024-01265-2

[24] Nipane A, Karmakar D, Kaushik N, Karande S, Lodha S. Few-layer MoS_2 p-type devices enabled by selective doping using low energy phosphorus implantation. ACS Nano. 2016;**10**:2128-2137. DOI: 10.1021/acsnano.5b06529

[25] Zeng Y, Yin S, Jin W, Hu Y, Wu S, Guo Z, et al. High-performance p-type MoS2 field-effect transistor by toroidal-magnetic-field controlled oxygen plasma doping. 2D Materials. 2019;**6**:025007. DOI: 10.1088/2053-1583/aafe2d

[26] Suleman M, Lee S, Kim M, Riaz M, Abbas Z, Park H, et al. Unveiling the potential of vanadium-doped CVD-grown p-type MoS_2 in vertical homojunction UV–vis photodiodes. Materials Today Physics. 2024;**43**:101427. DOI: 10.1016/j.mtphys.2024.101427

[27] Pető J, Ollár T, Vancsó P, Popov ZI, Magda GZ, Dobrik G, et al. Spontaneous doping of the basal plane of MoS_2 single layers through oxygen substitution under ambient conditions. Nature Chemistry. 2018;**10**:1246-1251. DOI: 10.1038/s41557-018-0136-2

[28] Li X, Xiong X, Li T, Li S, Zhang Z, Wu Y. Effect of dielectric Interface on the performance of MoS_2 transistors. ACS Applied Materials & Interfaces. 2017;**9**:44602-44608. DOI: 10.1021/acsami.7b14031

[29] Feng S, Tan J, Zhao S, Zhang S, Khan U, Tang L, et al. Synthesis of ultrahigh-quality monolayer molybdenum disulfide through in situ defect healing with thiol molecules. Small. 2020;**16**:2003357. DOI: 10.1002/smll.202003357

[30] Liu H, Ye PD. MoS_2 dual-gate MOSFET with atomic-layer-deposited Al_2O_3 as top-gate dielectric. IEEE Electron Device Letters. 2012;**33**:546-548. DOI: 10.1109/LED.2012.2184520

[31] Liu X, Qu D, Ryu J, Ahmed F, Yang Z, Lee D, et al. P-type polar transition of chemically doped multilayer MoS_2 transistor. Advanced Materials. 2016;**28**:2345-2351. DOI: 10.1002/adma.201505154

[32] Pradhan D, Gartia A, Sahoo KK, Ghosh SP, Kar JP. Fabrication of p-MoS_2/n-Si heterojunction by AuCl3 treatment for IR detection. Applied Physics A. 2022;**128**:248. DOI: 10.1007/s00339-022-05353-y

[33] Jang J, Kim J-K, Shin J, Kim J, Baek K-Y, Park J, et al. Reduced dopant-induced scattering in remote charge-transfer-doped MoS_2 field-effect transistors. Science Advances. 2025;**8**:eabn3181. DOI: 10.1126/sciadv.abn3181

[34] Zhang S, Le ST, Richter CA, Hacker CA. Improved contacts to p-type MoS_2 transistors by charge-transfer doping and contact engineering. Applied Physics Letters. 2019;**115**:073106. DOI: 10.1063/1.5100154

[35] Neal AT, Pachter R, Mou S. P-type conduction in two-dimensional MoS_2 via oxygen incorporation. Applied Physics Letters. 2017;**110**(19):193103. DOI: 10.1063/1.4983092

[36] Tang J, Wei Z, Wang Q, Wang Y, Han B, Li X, et al. In situ oxygen doping of monolayer MoS_2 for novel electronics. Small. 2020;**16**:2004276. DOI: 10.1002/smll.202004276

[37] Wei Z, Tang J, Li X, Chi Z, Wang Y, Wang Q, et al. Wafer-scale oxygen-doped MoS_2 monolayer. Small Methods. 2021;**5**:2100091. DOI: 10.1002/smtd.202100091

[38] Dolui K, Rungger I, Das Pemmaraju C, Sanvito S. Possible doping strategies for MoS_2 monolayers: An *ab initio* study. Physical Review B. 2013;**88**:075420. DOI: 10.1103/PhysRevB.88.075420

[39] Okada M, Nagamura N, Matsumura T, Ando Y, Lu AKA, Okada N, et al. Growth of MoS_2–Nb-doped MoS_2 lateral homojunctions: A monolayer p–n diode by substitutional doping. APL Materials. 2021;**9**:121115. DOI: 10.1063/5.0070333

[40] Zhang L, Wang Z, Zhang J, Chen B, Liang Z, Quan X, et al. Quasi-continuous tuning of carrier polarity in monolayered molybdenum dichalcogenides through substitutional vanadium doping. Advanced Functional Materials. 2022;**32**:2204760. DOI: 10.1002/adfm.202204760

[41] Li M, Wu X, Guo W, Liu Y, Xiao C, Ou T, et al. Controllable p-type doping of monolayer MoS_2 with tantalum by one-step chemical vapor deposition. Journal of Materials Chemistry C. 2022;**10**:7662-7673. DOI: 10.1039/D2TC01045C

[42] Laskar MR, Nath DN, Ma L, Lee EW II, Lee CH, Kent T, et al. p-type doping of MoS_2 thin films using Nb. Applied Physics Letters. 2014;**104**:092104. DOI: 10.1063/1.4867197

[43] Li M, Yao J, Wu X, Zhang S, Xing B, Niu X, et al. P-type doping in large-area monolayer MoS_2 by chemical vapor deposition. ACS Applied Materials and Interfaces. 2020;**12**:6276-6282. DOI: 10.1021/acsami.9b19864

[44] Li W, Huang J, Han B, Xie C, Huang X, Tian K, et al. Molten-salt-assisted chemical vapor deposition process for substitutional doping of monolayer MoS_2 and effectively altering the electronic structure and phononic properties. Advanced Science. 2020;**7**(16):2001080. DOI: 10.1002/advs.202001080

[45] Estrada CJ, Ma Z, Chan M. Complementary two-dimensional (2-D) MoS_2 FET technology. In: ESSDERC 2021 - IEEE 51st European Solid-State Device Research Conference (ESSDERC). 2021. pp. 219-222. DOI: 10.1109/ESSDERC53440.2021.9631795

[46] Wang J, Guo X, Yu Z, Ma Z, Liu Y, Lin Z, et al. Low-power complementary inverter with negative capacitance 2D semiconductor transistors. Advanced Functional Materials. 2020;**30**:2003859. DOI: 10.1002/adfm.202003859

[47] Chou A-S, Hsu C-H, Lin Y-T, Arutchelvan G, Chen E, Hung TYT, et al. Status and performance of integration modules toward scaled CMOS with transition metal dichalcogenide channel. In: 2023 International electron devices meeting (IEDM). San Francisco, CA, USA: IEEE; 2023. pp. 1-4. DOI: 10.1109/IEDM45741.2023.10413779

[48] Wang Y, Chhowalla M. Making clean electrical contacts on 2D transition metal dichalcogenides. Nature Reviews Physics. 2022;**4**:101-112. DOI: 10.1038/s42254-021-00389-0

[49] Banerjee S, Cao L, Ang YS, Ang LK, Zhang P. Reducing contact resistance in two-dimensional-material-based electrical contacts by roughness engineering. Physical Review Applied. 2020;**13**:64021. DOI: 10.1103/PhysRevApplied.13.064021

[50] Li W, Gong X, Yu Z, Ma L, Sun W, Gao S, et al. Approaching the quantum limit in two-dimensional semiconductor contacts. Nature. 2023;**613**:274-279. DOI: 10.1038/s41586-022-05431-4

[51] Batool S, Idrees M, Han S-T, Roy VAL, Zhou Y. Electrical contacts with 2D materials: Current developments and future prospects. Small. 2023;**19**:2206550. DOI: 10.1002/smll.202206550

[52] Ma N, Jena D. Charge scattering and mobility in atomically thin semiconductors. Physical Review X. 2014;**4**:11043. DOI: 10.1103/PhysRevX.4.011043

[53] Hosseini M, Elahi M, Pourfath M, Esseni D. Strain induced mobility modulation in single-layer MoS_2. Journal of Physics D: Applied Physics. 2015;**48**:375104. DOI: 10.1088/0022-3727/48/37/375104

[54] Wang J, Yao Q, Huang C-W, Zou X, Liao L, Chen S, et al. High mobility MoS_2 transistor with low Schottky barrier contact by using atomic thick h-BN as a tunneling layer. Advanced Materials. 2016;**28**:8302-8308. DOI: 10.1002/adma.201602757

[55] Park S, Schultz T, Shin D, Mutz N, Aljarb A, Kang HS, et al. The Schottky–Mott rule expanded for two-dimensional semiconductors: Influence of substrate

dielectric screening. ACS Nano. 2021;**15**:14794-14803. DOI: 10.1021/acsnano.1c04825

[56] Bian Z, Miao J, Zhang T, Chen H, Zhu Q, Chai J, et al. Carrier modulation in 2D transistors by inserting interfacial dielectric layer for area-efficient computation. Small. 2023;**19**:2206791. DOI: 10.1002/smll.202206791

[57] Lin C-Y, Simbulan KB, Hong C-J, Li K-S, Zhong Y-L, Su Y-K, et al. Polarity-controllable MoS_2 transistor for adjustable complementary logic inverter applications. Nanoscale Horizons. 2020;**5**:163-170. DOI: 10.1039/C9NH00275H

[58] Le OK, Chihaia V, Van On V, Son DN. N-type and p-type molecular doping on monolayer MoS_2. RSC Advances. 2021;**11**:8033-8041. DOI: 10.1039/D0RA10075G

[59] Park KH, Jo J-S, Choi J, Kim MJ, Chung K-B, Hong YK, et al. Electron donor or acceptor behavior of a $AuCl_3$ dopant manipulated by dip-pen nanolithography on a MoS2 thin-film transistor. Applied Surface Science. 2022;**588**:152846. DOI: 10.1016/j.apsusc.2022.152846

[60] Das S, Demarteau M, Roelofs A. Nb-doped single crystalline MoS_2 field effect transistor. Applied Physics Letters. 2015;**106**(17):173506. DOI: 10.1063/1.4919565

[61] Ma Z, Zhang L, Zhou C, Chan M. High current Nb-doped P-channel MoS_2 field-effect transistor using Pt contact. IEEE Electron Device Letters. 2021;**42**:343-346. DOI: 10.1109/LED.2021.3056178

[62] Sotthewes K, van Bremen R, Dollekamp E, Boulogne T, Nowakowski K, Kas D, et al. Universal fermi-level pinning in transition-metal dichalcogenides. The Journal of Physical Chemistry C. 2019;**123**:5411-5420. DOI: 10.1021/acs.jpcc.8b10971

[63] Gong C, Colombo L, Wallace RM, Cho K. The unusual mechanism of partial fermi level pinning at metal–MoS_2 interfaces. Nano Letters. 2014;**14**:1714-1720. DOI: 10.1021/nl403465v

[64] Kim C, Moon I, Lee D, Choi MS, Ahmed F, Nam S, et al. Fermi level pinning at electrical metal contacts of monolayer molybdenum dichalcogenides. ACS Nano. 2017;**11**:1588-1596. DOI: 10.1021/acsnano.6b07159

[65] Wang S, Robertson A, Warner JH. Atomic structure of defects and dopants in 2D layered transition metal dichalcogenides. Chemical Society Reviews. 2018;**47**:6764-6794. DOI: 10.1039/C8CS00236C

[66] Ko K, Jang M, Kwon J, Suh J. Native point defects in 2D transition metal dichalcogenides: A perspective bridging intrinsic physical properties and device applications. Journal of Applied Physics. 2024;**135**:100901. DOI: 10.1063/5.0185604

[67] Sorkin V, Zhou H, Yu ZG, Ang K-W, Zhang Y-W. The effects of point defect type, location, and density on the Schottky barrier height of Au/MoS_2 heterojunction: A first-principles study. Scientific Reports. 2022;**12**:18001. DOI: 10.1038/s41598-022-22913-7

[68] Liu X, Choi MS, Hwang E, Yoo WJ, Sun J. Fermi level pinning dependent 2D semiconductor devices: Challenges and prospects. Advanced Materials. 2022;**34**:2108425. DOI: 10.1002/adma.202108425

[69] Wang X, Hu Y, Kim SY, Cho K, Wallace RM. Mechanism of fermi

level pinning for metal contacts on molybdenum dichalcogenide. ACS Applied Materials and Interfaces. 2024;**16**:13258-13266. DOI: 10.1021/acsami.3c18332

[70] Schulman DS, Arnold AJ, Das S. Contact engineering for 2D materials and devices. Chemical Society Reviews. 2018;**47**:3037-3058. DOI: 10.1039/C7CS00828G

[71] Noori K, Xuan F, Quek SY. Origin of contact polarity at metal-2D transition metal dichalcogenide interfaces. npj 2D Materials and Applications. 2022;**6**:73. DOI: 10.1038/s41699-022-00349-x

[72] Song B, Gu H, Fang M, Guo Z, Ho Y-T, Chen X, et al. 2D niobium-doped MoS_2: Tuning the exciton transitions and potential applications. ACS Applied Electronic Materials. 2021;**3**:2564-2572. DOI: 10.1021/acsaelm.1c00121

[73] Chanana A, Mahapatra S. Theoretical insights to niobium-doped monolayer MoS_2–gold contact. IEEE Transactions on Electron Devices. 2015;**62**:2346-2351. DOI: 10.1109/TED.2015.2433931

[74] Kerelsky A, Nipane A, Edelberg D, Wang D, Zhou X, Motmaendadgar A, et al. Absence of a band gap at the interface of a metal and highly doped monolayer MoS_2. Nano Letters. 2017;**17**:5962-5968. DOI: 10.1021/acs.nanolett.7b01986

[75] Chuang S, Battaglia C, Azcatl A, McDonnell S, Kang JS, Yin X, et al. MoS_2 P-type transistors and diodes enabled by high work function MoO_x contacts. Nano Letters. 2014;**14**:1337-1342. DOI: 10.1021/nl4043505

[76] Yang Z, Kim C, Lee KY, Lee M, Appalakondaiah S, Ra C-H, et al. A fermi-level-pinning-free 1D electrical contact at the intrinsic 2D MoS_2–metal junction. Advanced Materials. 2019;**31**:1808231. DOI: 10.1002/adma.201808231

[77] Liu Y, Guo J, Zhu E, Liao L, Lee S-J, Ding M, et al. Approaching the Schottky–Mott limit in van der Waals metal–semiconductor junctions. Nature. 2018;**557**:696-700. DOI: 10.1038/s41586-018-0129-8

[78] Wang Y, Kim JC, Wu RJ, Martinez J, Song X, Yang J, et al. Van der Waals contacts between three-dimensional metals and two-dimensional semiconductors. Nature. 2019;**568**:70-74. DOI: 10.1038/s41586-019-1052-3

[79] Wang Y, Kim JC, Li Y, Ma KY, Hong S, Kim M, et al. P-type electrical contacts for 2D transition-metal dichalcogenides. Nature. 2022;**610**:61-66. DOI: 10.1038/s41586-022-05134-w

[80] Fang N, Toyoda S, Taniguchi T, Watanabe K, Nagashio K. Full energy spectra of interface state densities for *n*-and *p*-type MoS_2 field-effect transistors. Advanced Functional Materials. 2019;**29**(49):1904465. DOI: 10.1002/adfm.201904465

[81] Zheng W, Yuan B, Villena MA, Zhu K, Pazos S, Shen Y, et al. The origin and mitigation of defects induced by metal evaporation in 2D materials. Materials Science and Engineering: R: Reports. 2024;**160**:100831. DOI: 10.1016/j.mser.2024.100831

[82] Zhao Y, Tripathi M, Čerņevičs K, Avsar A, Ji HG, Gonzalez Marin JF, et al. Electrical spectroscopy of defect states and their hybridization in monolayer MoS_2. Nature Communications. 2023;**14**:44. DOI: 10.1038/s41467-022-35651-1

[83] Gao L, Zhang X, Yu H, Hong M, Wei X, Chen Z, et al. Deciphering

vacancy defect evolution of 2D MoS_2 for reliable transistors. ACS Applied Materials & Interfaces. 2023;**15**:38603-38611. DOI: 10.1021/acsami.3c07806

[84] Lan Y-W, Chen P-C, Lin Y-Y, Li M-Y, Li L-J, Tu Y-L, et al. Scalable fabrication of a complementary logic inverter based on MoS_2 fin-shaped field effect transistors. Nanoscale Horizons. 2019;**4**:683-688. DOI: 10.1039/C8NH00419F

[85] Ma Z, Estrada CJ, Gong K, Zhang L, Chan M. On-chip integrated high gain complementary MoS_2 inverter circuit with exceptional high hole current P-channel field-effect transistors. Advanced Electronic Materials. 2022;**8**:2200480. DOI: 10.1002/aelm.202200480

[86] Min S-W, Yoon M, Yang SJ, Ko KR, Im S. Charge-transfer-induced p-type channel in MoS_2 flake field effect transistors. ACS Applied Materials & Interfaces. 2018;**10**:4206-4212. DOI: 10.1021/acsami.7b15863

[87] Guo Y, Li J, Zhan X, Wang C, Li M, Zhang B, et al. Van der Waals polarity-engineered 3D integration of 2D complementary logic. Nature. 2024;**630**:346-352. DOI: 10.1038/s41586-024-07438-5

[88] Liang M, Yan H, Wazir N, Zhou C, Ma Z. Two-dimensional semiconductors for state-of-the-art complementary field-effect transistors and integrated circuits. Nanomaterials. 2024;**14**:1408. DOI: 10.3390/nano14171408

[89] Tong L, Wan J, Xiao K, Liu J, Ma J, Guo X, et al. Heterogeneous complementary field-effect transistors based on silicon and molybdenum disulfide. Nature Electronics. 2023;**6**:37-44. DOI: 10.1038/s41928-022-00881-0

[90] Estrada CJ, Ma Z, Chan M. Complementary two-dimensional (2-D) FET technology with MoS_2/hBN/graphene stack. IEEE Electron Device Letters. 2021;**42**:1890-1893. DOI: 10.1109/LED.2021.3124823

[91] Knobloch T, Illarionov YY, Ducry F, Schleich C, Wachter S, Watanabe K, et al. The performance limits of hexagonal boron nitride as an insulator for scaled CMOS devices based on two-dimensional materials. Nature Electronics. 2021;**4**:98-108. DOI: 10.1038/s41928-020-00529-x

[92] Lee D, Lee JJ, Kim YS, Kim YH, Kim JC, Huh W, et al. Remote modulation doping in van der Waals heterostructure transistors. Nature Electronics. 2021;**4**:664-670. DOI: 10.1038/s41928-021-00641-6

[93] Li H-M, Lee D, Qu D, Liu X, Ryu J, Seabaugh A, et al. Ultimate thin vertical p–n junction composed of two-dimensional layered molybdenum disulfide. Nature Communications. 2015;**6**:6564. DOI: 10.1038/ncomms7564

Chapter 5

MoS_2 Exfoliated in Liquid Phase: Size Selection, Networks, and Applications

Daniel Olaya-Cortes and Yenny Hernandez

Abstract

Molybdenum disulfide (MoS_2) is the most extensively studied transition metal dichalcogenide (TMD), with many of its significant characteristics first identified in this material. This chapter presents the method to produce dispersions of MoS_2 using liquid phase exfoliation (LPE). LPE produces nanosheets with wide distributions of dimensions, and by using engineered centrifugation techniques, researchers can select the size of the dispersions depending on the required application. By printing these dispersions, the nanosheets are arranged through networks, and the properties of the system now depend on the synergy between the intrinsic properties of the nanosheets and the junctions. Finally, we show applications of MoS_2 networks in the fields of sensing, energy conversion, charge storage, and non-linear optics.

Keywords: liquid phase exfoliation, size selection, transport mechanism, junction-limited conductivity, networks

1. Introduction

Molybdenum disulfide (MoS_2) is the most studied transition metal dichalcogenide (TMD) [1–4]. This family of materials has been studied as they can be conductors, semiconductors, topological insulators, and even superconductors. As described by their name, their chemical formula is of the form MC_2, where M is a transition metal atom, such as molybdenum or tungsten, and C corresponds to a chalcogen, such as sulfur, selenide, or tellurium [1, 5]. The most valuable behavior of these materials is their transition from an indirect-gap semiconductor to a direct one as they are exfoliated to their monolayer version, which opens up their presence in applications in photodetectors, saturable absorbers, solar cells, transistors, and valleytronics [2].

These materials are produced using either bottom-up or top-down techniques [3]. However, among these approaches, liquid phase exfoliation (LPE) [4] is preferred for its scalability potential. This process involves applying energy to bulk material within a liquid medium to induce delamination. LPE encompasses various methods, including probe sonication, electrochemical exfoliation, intercalation, and shear exfoliation [6–10]. The second section of this chapter will discuss these techniques in detail, with particular emphasis on solvent selection and the most commonly used solvents in LPE [11].

LPE typically produces dispersions of nanosheets with a wide distribution of lateral dimensions and thicknesses [4]. However, centrifugation-based techniques have been employed to refine these distributions and obtain nanosheets with specific areas and thicknesses. This can be achieved through either liquid cascade centrifugation [12] or density-gradient centrifugation [13]. The third section of this chapter provides a comparative analysis of these two methods, examining how researchers use dimension-dependent physical properties to analyze distributions and establish useful characterization metrics.

Once exfoliated, nanosheets can be used for fundamental research as well as applied science. Among their applications, they can be assembled into networks, either by printing to form films or by mixing with polymers to create composites [14–16]. Given that LPE produces nanosheets with a broad range of dimensions, different fabrication approaches can be tailored to specific applications. This chapter focuses on printing techniques and polymer composites; thus, in the fourth section, readers will find a summary of the most commonly used methods, along with a comparison based on film thickness, nanosheet lateral size, and electrical conductivity. Additionally, a brief discussion on the transport characteristics of such networks is provided.

To conclude this chapter, we provide an overview of the latest and more successful implementations of MoS_2 networks in applied physics. As previously discussed, band structure modifications in MoS_2 enable its application across various fields, including sensing [17], energy storage [18], energy conversion [19], computing [20], and nonlinear optics [21]. The final section of this chapter presents a summary of these applications, with a table highlighting key performance parameters, facilitating comparison across different devices.

2. Liquid phase exfoliation

Nanosheets of MoS_2 could be produced by either bottom-up or top-down techniques [3, 4, 22, 23]. The main difference between both techniques is the initial form of the reactants. Bottom-up approaches involve the careful modification of atoms to form nanostructures. Its main advantage is the production of high-quality monolayers that allowed fundamental research and the discovery of novel characteristics on those nanosheets. However, these approaches suffer from low yield, small lateral size nanosheets, and additional chemical steps to obtain them in an isolated form [3].

In contrast with bottom-up techniques, top-down approaches involve breaking down layered materials to obtain few-layer and monolayer nanosheets. Yet, in these approaches, methods such as mechanical exfoliation also yield high-quality monolayers, however at a low production rate, posing a challenge for scale-up prospects [22]. Liquid phase exfoliation (LPE) groups all methods where energy is applied in a liquid medium to break interlayer attractions in such layered materials and to avoid aggregation of the nanosheets [4, 23].

LPE can be achieved through various methods, including electrochemical exfoliation, intercalation (also known as chemical exfoliation), probe and bath sonication, and shear exfoliation. Furthermore, careful attention is required in LPE methods when considering the solvent used and the type of energy transfer within the liquid medium [4, 6, 7, 23].

2.1 Solvent choice

Molybdenum disulfide has been properly exfoliated in the solvents as shown in **Figure 1A**. Solvent selection requires the correct match of the surface tension of the solvent and the surface energy of the material to exfoliate and obtain nanosheets as observed in **Figure 1B** [4, 23]. This selection is done to ensure that the solvent effectively transfers acoustic power during sonication, that the enthalpy of mixing between the solvent and the interlayer faces of the laminar crystal is diminished, and that the solvent stabilizes the dispersion of nanosheets by providing a steric barrier to prevent aggregation [4].

Hansen solubility parameters have been used to aid this selection. These are the dispersion, polar and hydrogen-bonding component of the Hildebrand parameter, which is related to the cohesive energy of the solvent [11]. Solvents with Hansen

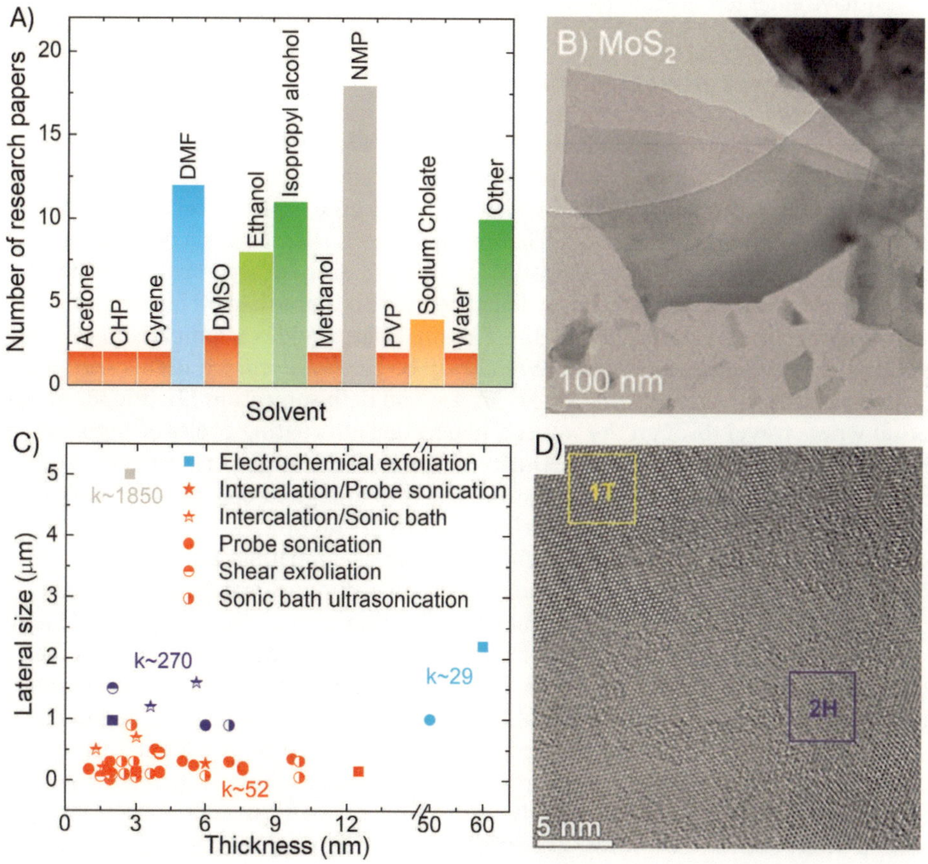

Figure 1.
(A) Histogram of the most common solvents used to exfoliate MoS_2 in liquid phase. NMP, DMF, and isopropyl alcohol (IPA) are the most used, while the other alternatives are ammonia [19], aqueous 1-pyrenesulfonic acid sodium salt (PS1) [24], chloroform and 1-dodecanethiol [25], dimethyl sulfoxide (DMSO) [11], ethylene glycol [26], polyvinylpyrrolidone (PVP) [27], sodium deoxycholate (SDC) [21], sugar-based solvents [28], and tetrahydrofuran (THF) [29]. (B) Representative scanning electron microscope (SEM) figure of an MoS_2 nanosheet. Figure adapted with permission from Ref. [4] Copyright 2011 Science Publishing Group. (C) Cluster analysis of nanosheet dimensions of MoS_2 exfoliated in liquid phase through different liquid phase exfoliation methods. (D) High-resolution transmission electron microscopy (HR-TEM) of an MoS_2 nanosheet displaying regions of both the metallic 1 T-phase and the semiconductor 2H-phase. Figure adapted with permission from Ref. [7] Copyright 2018 Elsevier B.V.

solubility parameters closely aligned to those of TMD nanosheets are effective in facilitating their exfoliation. Values for MoS_2 and other TMDs are $\delta_D \approx 18$ MPa$^{1/2}$, $\delta_P \approx 8.5$ MPa$^{1/2}$, and $\delta_H \approx 7$ MPa$^{1/2}$. Furthermore, exfoliation efficiency and size selection yields can be affected by viscosity, with many effective solvents exhibiting a viscosity higher than 1 mPa·s [30].

Commonly used solvents for effective exfoliation are nitrogen-functionalized liquids, such as N-cyclohexyl pyrrolidone (CHP) and N-methylpyrrolidone (NMP) [8, 9, 11, 14, 29, 31–37]. Among these, NMP is the most widely utilized, as shown in **Figure 1A**, due to its ability to produce high-concentration dispersions that remain stable for several months, which is an essential factor for scalability. However, both NMP and dimethylformamide (DMF) have high boiling points and pose health and environmental hazards [11, 26, 27, 38–41], which limit their suitability for large-scale applications. Thus, greener alternatives such as ethanol [7, 10, 26, 40] and isopropyl alcohol (IPA) [11, 18, 37, 40, 42, 43] have emerged as effective substitutes, as indicated in **Figure 1A**.

2.2 Types of liquid phase exfoliation

Clustering analysis was conducted, as shown in **Figure 1C**, to evaluate nanosheet dimensions across different liquid phase exfoliation methods. The results reveal four distinct groups: two primary categories classified as high aspect ratio (270) and low aspect ratio (52), along with two outlier groups. One outlier consists of nanosheets with exceptionally large aspect ratios [35], corresponding to large-area nanosheets, while the other includes two pairs of nanosheets with more than 50 layers and a notably small aspect ratio of 29 [18, 32]. This analysis provides a comparison of various liquid phase exfoliation techniques, which will be briefly discussed in this section.

In both probe [14, 15, 19, 29, 31, 32, 37, 43] and bath sonication [21, 26, 38, 41, 44], sound waves travel through the solvent, generating alternating cycles of high and low pressure that facilitate exfoliation. During these cycles, energy is transferred to the material through vibration and cavitation, the formation and collapse of bubbles, which progressively thins the material until a dispersion of nanosheets can be obtained through post-processing techniques such as centrifugation. Moreover, the extent of exfoliation and the dimensions of the two-dimensional (2D) flakes are influenced by parameters such as sonication duration, power, and the physicochemical properties of the solvent. In probe sonication, energy is directly applied to the dispersion medium [14, 15, 19, 29, 31, 32, 37, 43], while in bath sonication, it must travel through the water bath and vial before reaching the sample [21, 26, 38, 41, 44].

As illustrated in **Figure 1C**, both probe and bath sonication predominantly yield low aspect ratio nanosheets; however, a few studies have demonstrated the production of high aspect ratio nanosheets through a pretreatment process using NMP [41] and one study shows notably low aspect ratio when centrifugation is skipped from this techniques [32]. These sonication methods are advantageous due to their simplicity, non-oxidative nature, and ability to produce high-quality nanosheets. However, they also present limitations, including the use of environmentally harmful solvents, low to medium yields, and the generation of relatively small nanosheets [31, 43, 44].

Intercalation, also known as chemical exfoliation, is a method in which chemical species are incorporated into the interlayer cavity of the layered material, weakening van der Waals forces between layers and allowing its separation [7–9, 20, 36, 45, 46]. Subsequent sonication is sufficient to produce dispersions of MoS_2 nanosheets. During intercalation-assisted sonication in water, lithium hydroxide (LiOH) and hydrogen

(H_2) gas bubbles form, facilitating the exfoliation of the layered material. Numerous studies have demonstrated that lithium ion is the most effective intercalant for bulk TMDs [7–9, 45, 46]. Additionally, other intercalants and chemical modification techniques have been explored [20, 36] There is ongoing interest in intercalating small molecules capable of expanding within the TMD interlayers upon activation, through either decomposition or other mechanisms.

Although this process mostly produces high aspect ratio nanosheets, as shown in **Figure 1C**, it requires harsh conditions such as inert atmospheres, long reaction times, and some exfoliation products to undergo modifications in their electronic structure. Specifically, the intercalation of bulk MoS_2 is known to induce a transition of the semiconducting 2H-phase to the metallic 1 T-phase [7–9, 20, 45], as shown in **Figure 1D**. This method allows to produce dispersions of 1 T-MoS_2, which are stable for months. However, this structural phase is instable and the sheets convert back to the 2H-phase in a matter of days, which can be sped up by annealing at 400° in an inert atmosphere [7, 20, 45].

Electrochemical exfoliation involves inserting ions into the interlayer spaces of a material by applying a constant potential or current. Unlike conventional intercalation, where ions are introduced without an external electric potential, electrochemical exfoliation relies on an applied voltage to facilitate ion insertion [6, 18, 35, 39, 47]. Depending on the polarity of the applied voltage, exfoliation can be either anodic or cathodic. Anodic exfoliation occurs in aqueous electrolytes containing acids or inorganic salts [6, 18, 35, 47], while cathodic exfoliation takes place in organic solvents with lithium or alkylammonium salts as electrolytes [39].

As illustrated in **Figure 1C**, this exfoliation method predominantly yields low aspect ratio nanosheets having a few cases of high aspect ratio nanosheets [27]. However, it also produces nanosheets that fall within the outlier categories of exceptionally large and notably small aspect ratio. Regarding the formation of large-area nanosheets, Liu and coworkers attribute this outcome to the advantages of the method, including a low degree of oxidation, ease of control, and scalability [35]. In contrast, Zhuo and coworkers utilized this technique to exfoliate and functionalize MoS_2, yet the resulting nanosheets contained more than 50 layers, which is a result likely influenced by the choice of intercalant [18]. The intercalation of large cations induces mechanical strain, which obstructs further intercalation into adjacent layers and ultimately prevents full exfoliation.

Finally, shear exfoliation entails applying shear forces to delaminate nanosheets from a layered crystal within a liquid medium. A critical parameter in this process is the shear rate, which must exceed 10^4 s^{-1} [10, 28, 34, 40]. Commonly used solvents include IPA and NMP, both of which facilitate exfoliation and help stabilize the resulting nanosheets [34, 40]. As shown in **Figure 1C**, most nanosheets produced via this technique exhibit a low aspect ratio. However, Yuan and coworkers successfully obtained nanosheets with lateral dimensions of 1 μm when using bulk powders with lateral sizes of up to 40 μm [10]. The key advantages of this method include its scalability and low cost, being environmentally friendly, and having non-oxidative nature. However, it tends to produce inhomogeneous flakes [10, 28, 34, 40].

3. Size selection

This section focuses on size selection, a method developed by Backes and colleagues [12], in which liquid cascade centrifugation (LCC) is used to control the dimensions of nanosheets. In this technique, centrifugation is performed at increasing speeds, producing dispersions where nanosheets are separated based on size and thickness. This approach offers two key advantages. First, the sequential centrifugation steps, or cascade, can be tailored to specific applications, allowing for the isolation of nanosheets with dimensions suited to particular requirements. Second, this method facilitates the efficient production of dispersions containing nanosheets of varying lengths and thicknesses [12].

Insights can be drawn from the research of five distinct groups, as summarized in **Figure 2A** through **D**, with each group represented by a unique symbol [8, 9, 12, 17, 42, 48]. **Figure 2A** illustrates a decrease in nanosheet dimensions as centrifugation speed increases. This is typically quantified by calculating the average of distributions obtained from scanning electron microscopy (SEM) for lateral dimensions and atomic force microscopy (AFM) for thickness measurements, as illustrated in **Figure 2B**. Additionally, the data reveal three distinct trends in nanosheet dimensions: high aspect ratio, medium aspect ratio, and low aspect ratio nanosheets. High aspect ratio nanosheets were obtained through bath sonication in NMP, assisted by lithium intercalation [8, 9]. Medium aspect ratio nanosheets were produced via bath sonication in

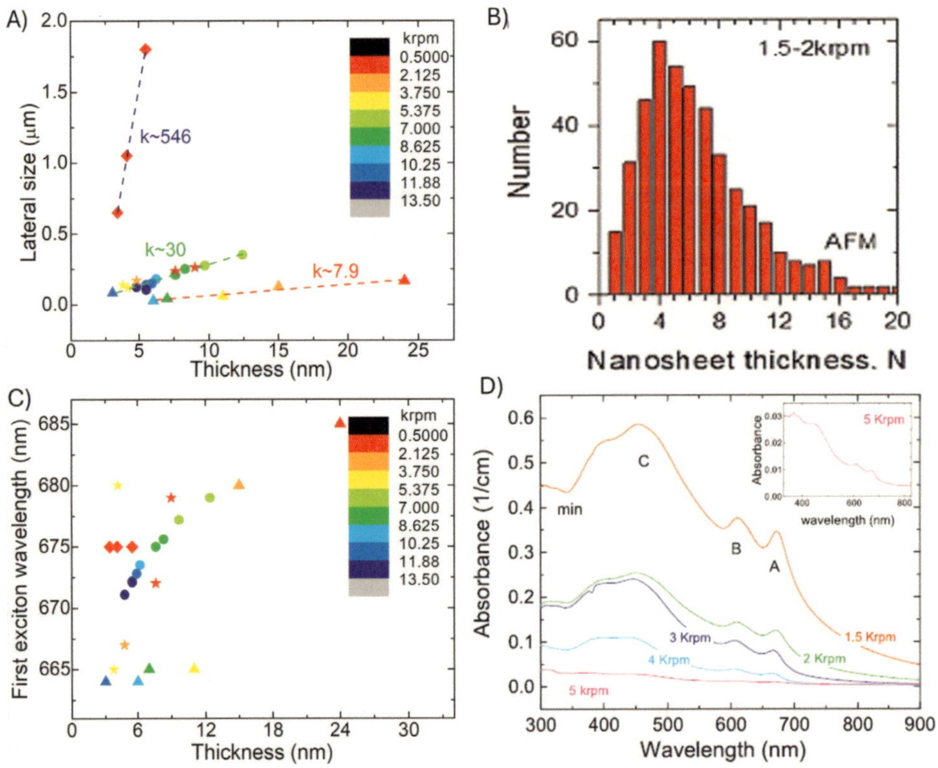

Figure 2.
(A) Dimensions of size-selected nanosheets through the method called liquid cascade centrifugation (LCC) as a function of centrifuge speed. Diamonds [8, 9], circles [42], stars [17], and triangles [48] correspond to the four groups that were analyzed in this section. (B) Representative histogram of thickness measured as N, number of monolayers per nanosheet, of a size-selected dispersion at a centrifuge speed of 1.5 krpm. Figure adapted with permission from Ref. [12] Copyright 2016 American Chemical Society. (C) Shift of the first exciton wavelength of the size-selected nanosheets as a function of centrifuge speed. (D) Extinction spectra of molybdenum disulfide dispersions as a function of centrifuge speed. Figure adapted with permission from Ref. [17] Copyright 2022 Elsevier B.V.

isopropyl alcohol (IPA) [42] and probe sonication in sodium cholate [17]. Low aspect ratio nanosheets were generated through bath sonication in sodium cholate [48].

Furthermore, **Figure 2C** and **D** demonstrate that as the speed of centrifugation increases, the exciton peak shifts toward lower wavelengths. This phenomenon can be attributed to the findings of Backes, who observed that higher centrifugation results in an increased concentration of monolayers in the dispersions [12]. These monolayers exhibit excitonic peaks at shorter wavelengths. Particular attention should be given to the study by Olaya and colleagues, which reports large-area MoS_2 nanosheets compared to other size-selected samples and employs centrifugation speeds below 2 krpm [8]. In this case, the excitonic peak does not exhibit the same shift, likely due to the fact that the nanosheets contain more than six layers, resulting in a lower monolayer content and a diminished excitonic response of such monolayers. This observation is significant, as the optical properties of nanosheets are highly dependent on the number of layers. Consequently, excitonic peak shifts have been used as a metric to estimate both monolayer content and the thickness of the nanosheets [49].

Although LCC enables the selection of nanosheets with specific dimensions [50], achieving uniform thickness remains a challenge. This issue is effectively addressed through density gradient ultracentrifugation (DGU), a technique developed by Hersam and coworkers. DGU utilizes ultracentrifugation to separate nanosheets based on their buoyant density, which is exclusively dependent on thickness [13]. However, molybdenum disulfide dispersed with sodium cholate exhibits a relatively high buoyant density, limiting the effectiveness of thickness-based separation. To overcome this limitation, the use of Pluronic F68, a block copolymer, reduces the buoyant density of MoS_2 nanosheets, thereby enabling efficient thickness sorting [13]. Furthermore, atomic force microscopy (AFM) measurements confirmed that different fractions obtained after DGU exhibit distinct average thicknesses. In contrast, the lateral size distribution showed significant overlap across fractions, indicating that the separation was primarily governed by thickness rather than lateral dimensions [13].

4. Networks

This section focuses on the research by Jonathan Coleman's group, which is a testament to the vast potential of MoS_2, graphene, and other two-dimensional (2D) materials. Their research on networks composed of different 2D materials highlights the adaptability and potential of these structures [16, 51]. Networks are composed of uniform structures extending over length scales greater than 10 μm. While maintaining overall uniformity, they exhibit local disorder while preserving strong connectivity, porosity, and in-plane alignment. These networks can be assembled using a diverse range of materials, including conducting nanosheets such as graphene and silver nanowires [52], semiconducting TMDs [14], and insulating materials like boron nitride [53].

Figure 3A presents a summary of the film thickness of molybdenum disulfide networks fabricated using various techniques and nanosheets with different average lateral dimensions. Common methods for producing such networks include ink-jet printing (IP) [33, 38], Langmuir-Blodgett (LB) [32, 54], polymer composite formation (PC) [15, 21], screen printing (SP) [55], spin coating [19, 20], spray coating (Spr) [14, 37], and vacuum filtration (VF) [8, 9, 18]. The resulting film thickness varies significantly, ranging from 10 nm to 100 μm. Ink-jet printing, spin coating, and spray coating can yield films with thicknesses below 300 nm [19, 20, 37, 38], while IP, LB,

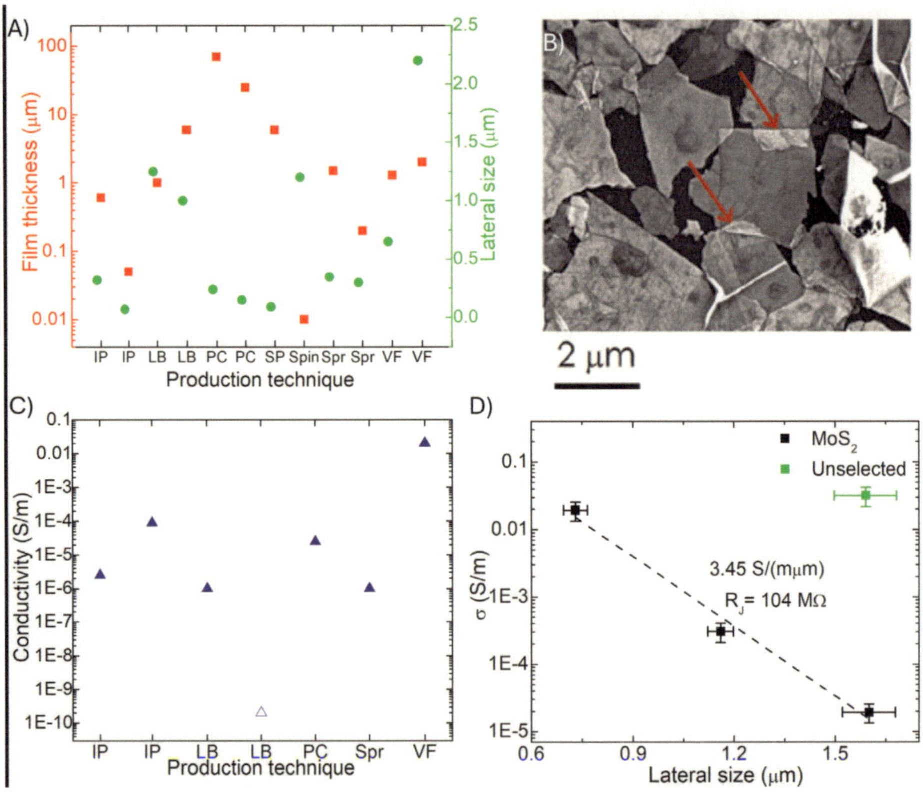

Figure 3.
(A) Film thickness (red axis) and average lateral size (green axis) of MoS_2 nanosheets as a function of the production technique. (B) Surface SEM figure of an electrochemically exfoliated MoS_2 network showing the junctions between nanosheets through the red arrows. Figure adapted with permission from Ref. [16] Copyright 2024 Nature Publishing Group. (C) Electrical conductivity of the nanosheet networks as a function of the production technique. Filled triangles correspond to measurements in-plane and the empty triangle corresponds to the out-of-plane conductivity. (D) In-plane conductivity as a function of lateral size. The model allows to find the junction resistance R_J. Figure adapted with permission from Ref. [9] Copyright 2025 IOP Publishing Ltd.

SP, Spr, and VF methods produce films within the range of 800 nm to 7 μm [8, 9, 14, 18, 32, 33, 54, 55]. In contrast, MoS_2-polymer composites can reach thicknesses exceeding 30 μm [15, 21].

Additionally, **Figure 3A** indicates that ink-jet printing, spray coating, and the formation of polymer composites are compatible with nanosheets smaller than 400 nm in lateral dimensions [14, 15, 21, 33, 37, 38]. Meanwhile, Langmuir-Blodgett assembly and spin coating are suitable for medium-size nanosheets [19, 20, 32, 54]. Notably, screen printing and vacuum filtration incorporate both small and large nanosheets, demonstrating the versatility of these approaches [8, 9, 55].

Ink-jet printing (IP) enables the precise patterning of films within complex and spatially confined structures. Due to its scalability potential, this technique offers advantages such as low cost, minimal material waste, and suitability for fabricating heterostructures [33, 38]. However, successful implementation requires careful ink formulation with a high nanosheet concentration. Additionally, nozzle clogging may occur due to the nanosheet size, and the presence of binders can reduce conductivity, often requiring post-processing treatments to enhance electrical performance [33, 38].

Spin coating is a widely used and well-established technique in the electronics industry due to its simplicity and reliability. This method provides precise control over film thickness and surface topography and can be utilized to prepare individual flakes for fundamental studies [19, 20]. Yet, similar to IP, spin coating suffers from high material waste, as a significant portion of the ink is discarded. Additionally, achieving homogeneous coverage requires high-concentration inks to prevent the formation of nonuniform films with overlapping flakes [19, 20, 33, 38].

Highly uniform films over large areas on a wide range of substrates using high concentration inks are fabricated through spray coating [14, 37]. However, this method lacks precision, often requiring masking techniques for patterned deposition, which increases material waste. Additionally, the resulting films may contain overlapping flakes, and the process demands careful optimization of droplet formation to ensure uniformity [14, 37]. Furthermore, IP, spin, and spray coating commonly rely on high boiling point solvents, which can reduce network conductivity [14, 19, 20, 33, 37, 38].

Langmuir-Blodgett assembly is a highly controllable film deposition technique that produces highly ordered structures with excellent reproducibility, precise control over packing density, and minimal material waste. However, it is a time-consuming process that requires high-purity components to ensure uniformity [32, 54]. Similarly, vacuum filtration is a simple and cost-effective method that minimizes residual solvents and film nonuniformity while providing control over film composition and deposited mass [8, 9, 18]. However, its application is primarily limited to initial characterization, as the filter size constrains geometric features. Both methods lack scalability, particularly for applications that require complex structural patterns or large-area fabrication [8, 9, 18, 32, 54].

Screen printing is a rapid, cost-effective manufacturing process that offers high reproducibility and the flexibility to create various structures, including large area, on both rigid and flexible substrates [55]. However, this technique has certain limitations, including the need for precise viscosity control of the ink. Additionally, achieving an optimal ratio between binders and MoS_2 is crucial to ensuring proper adhesion and cohesion within the printed structures [55].

Finally, polymer composite formation is a straightforward method that involves mixing dispersions to create composites and hybrid films. The polymer plays a crucial role in stabilizing nanosheets and modulating device properties [15, 21]. However, its incorporation requires careful evaluation, as polymers are typically insulating materials, which can pose challenges for electronic applications [15, 21].

The electrical conductivity of nanosheet networks arises from a complex interaction between the intrinsic properties of individual nanosheets and the overall collective behavior of the network [51]. These networks consist of nanosheets linked at junctions (refer to **Figure 3B**), where they are closely spaced and separated by a narrow van der Waals gap [16]. The conductivity of the networks is influenced not only by the intrinsic conductivity of the nanosheets but also by charge transport across their junctions. **Figure 3C** shows the wide range of electrical conductivity of the network as a function of production techniques.

Various transport regimes can be determined by the network's morphology. For example, depending on the degree of alignment within the network, mobility values can range from 0.1 to 10 cm^2/Vs [51]. Oblique alignments restrict mobility due to limited contact between junctions, leading to junction-limited conductivity [38]. In contrast, edge-to-edge alignments with densely packed nanosheets result in intermediate mobility levels [27]. Basal-plane alignments, where nanosheets maximize contact area and conform to each other, significantly enhance network mobility [20]. In this configuration, the network's conductivity approaches that of individual nanosheets due to the reduced junction resistance.

Research has also focused on enhancing the conductivity of nanosheet networks. For example, thermal and photonic annealing eliminates residual polymers and solvents from the dispersion, which can obstruct electrical conductivity. Additionally, applying pressure to the network decreases porosity and improves connectivity, leading to better nanosheet alignment, a key factor in facilitating efficient charge transport [51].

Further insights indicate that conductivity is inversely proportional to nanosheet thickness [56]. Monolayer and few-layer nanosheets, due to their low bending rigidity, can conform more easily to one another, thereby improving conductivity. Moreover, conductivity initially increases with nanosheet length but begins to decline after reaching a maximum [16]. Gabbett and Coleman developed a model that accounts for both the intrinsic properties of the nanosheets and the behavior of junction resistance. By applying this model to 2D material networks, researchers have successfully fitted the experimental data and extracted the value for the junction resistance, as shown in **Figure 3D** [9, 16].

Finally, the transport mechanisms in nanosheet networks are varied and temperature-dependent [51, 57]. At near-ambient and elevated temperatures, conduction follows an Arrhenius-like behavior, allowing the activation energy to be determined through model fitting. In certain networks, the activation network arises from a combination of thermal carrier activation and interparticle hopping [57]. The results support a model in which network conductivity is influenced equally by both nanosheet conductivity and junction transport. At low temperatures, however, tunneling becomes the dominant transport mechanism [51].

5. Applications

The last section of this chapter is dedicated to the applications of networks of molybdenum disulfide and their physical properties. **Table 1** gives a few recent applications of these types of devices. Zhuo and coworkers demonstrated that MoS_2 can undergo simultaneous electrochemical exfoliation and functionalization. Additionally, using electrochemical impedance spectroscopy, they found that the specific capacitance of MoS_2-PTFE–based electrodes increases by 25% upon functionalization [18].

Giovanni Neri's group tested drop-casted molybdenum disulfide on screen-printed carbon electrodes for biosensing applications. Specifically, they performed liquid cascade centrifugation (LCC) to test whether the dimensions would play a role in the sensing performance of dopamine (DA) and riboflavin (RF). DA performance was found to be independent of the dimensions of the nanosheets, while RF sensing was promoted in small lateral size and a few number of layers of MoS_2 [17]. Moreover, the researchers managed to sense both biomolecules in the linear range shown in the **Table 1**. Furthermore, Pavol Gemeiner's group also managed to sense DA with a larger linear range with screen-printed devices of high-viscosity inks made of both large and small MoS_2 particles [55].

Li and colleagues demonstrated that ink-jet–printed MoS_2 was integrated with printed silver nanoparticles to produce MoS_2-based thin-film photodetectors and demonstrate notable performance, confirming that MoS_2 retains its intrinsic properties after the printing process [38]. Another contribution by Coleman's group established that due to the relationship between dark conductivity and bandgap,

DOI: http://dx.doi.org/10.5772/intechopen.1010569

Application	Key parameter	Value	Lateral size (nm)	Ref
Electrochemical cell	Specific capacitance (F/cm^3)	10–20	2200	[18]
Electrochemical sensor	Linear range (μM)	10–100	120–260	[17]
		1–300	90–6000	[55]
Photodetector	Responsivity (μA/W)	0.2–1.5	650–1800	[8]
		1.5–4	105–350	[42]
		36	70	[38]
		10	1000	[54]
Piezoresistivity sensor	Gauge factor	−15–22	650–1600	[9]
		−25	240	[15]
Saturable absorber	Saturable intensity (MW/cm^2)	1.6	150	[21]
Solar cell	PCE (%)	14.9	140	[19]
Transistor	Mobility (cm^2/Vs)	0.73	975	[27]
		10	1200	[20]
		0.0001	300	[37]
		0.15	348[1]	[14]

1 Porosity was measured at a value of 43%.

Table 1.
Recent instances of the most common applications where MoS networks are used.

wide-bandgap TMDs such as MoS_2 and tungsten disulfide (WS_2) exhibit a much higher photo-to-dark conductivity ratio, despite displaying relatively low photoconductivity [54].

We also contributed to this field by finding that vacuum-filtered size-selected MoS_2-based devices showed higher responsivity as compared with WS_2-based devices and that the response time increases with lateral size [8]. Moreover, Alzakia and colleagues successfully printed a TMD-graphene heterostructure and found good photoresponse with acceptable rise and decay times showing good reproducibility across multiple devices [42].

We observed a transition from a negative piezoresistive gauge factor (GF) to a positive one in MoS_2 networks that occurs when compression is applied and as the lateral size increases. For tensile strain, the gauge factor shows positive values [9]. In contrast, Biccai and colleagues demonstrated that the intrinsic negative piezoresistance of MoS_2-polyethylene oxide nanocomposites causes a decrease in composite resistance when subjected to small tensile strains. However, at higher strain levels, resistance increases due to a rise in junction resistance [15]. The incorporation of nanosheets also led to composite reinforcement, with stiffness more than doubling at approximately 0.5 vol% MoS_2, suggesting effective stress transfer between the polymer and nanosheets [15].

Popov and Taylor's group successfully produced a few-layer MoS_2 polymer composite saturable absorber. They attribute sub-bandgap saturable absorption in MoS_2 to the presence of edge states within the material bandgap that arise due to the boundaries of a finite crystal structure [21]. Moreover, Anthopoulos' group used MoS_2, which was LPE using aqueous ammonia as a solvent, as the hole-transport layer material in an organic solar cell achieving a power conversion efficiency (PCE) of 14.9% [19].

Neilson and colleagues demonstrated the fabrication of densely packed, predominantly single-layer MoS_2 films by assembling flakes at the interface of two immiscible liquids, driven by interfacial tension gradients. A field-effect transistor (FET) based on these films exhibited a peak mobility of 0.73 $cm^2/(Vs)$ and an on/off ratio of 10^5, comparable to FETs made from chemical vapor deposition (CVD)-grown MoS_2 and mechanically cleaved few-layer MoS_2 single crystals using a back-gate geometry with silicon dioxide (SiO_2) as the gate dielectric [27].

Duan and colleagues successfully achieved electrochemical intercalation of quaternary ammonium molecules into MoS_2, followed by sonication to exfoliate the layers. The resulting nanosheets were then used to fabricate high-performance thin-film transistors, where basal-plane alignment optimized charge transport across the stacked nanosheets. These devices exhibited room-temperature mobilities of approximately 10 $cm^2/(Vs)$ and on/off ratios of 10^6, significantly surpassing the performance of previously reported solution-processed MoS_2 transistors [20].

To analyze charge transport, bottom-gate thin-film transistors utilizing exfoliated MoS_2 films were fabricated through spray coating by Wagner's group. Electrical measurements revealed that applying a polymethyl methacrylate (PMMA) layer over the semiconductor significantly enhanced the device's electrical performance. The evaluation of the electrical properties indicates that the PMMA layer enhances charge transfer between adjacent flakes [37]. Further enhancement can be done by electrolytic gating using an ionic liquid within the networks done by Coleman's group. The transconductance, defined as the gradient of the source-drain current with respect to the gate voltage, was found to scale with the thickness of the TMD networks [14].

6. Conclusions

This chapter explores the key aspects of liquid phase exfoliation (LPE) of MoS_2, focusing on its use in networks and polymer composites, which have been successfully applied in various areas of condensed matter physics. The discussion begins with an overview of the main advantages and limitations of LPE techniques. Additionally, these methods are classified into four distinct categories based on the resulting nanosheet dimensions.

The following section of this chapter examines two distinct size-selection processes designed to refine the broad size distribution inherent to LPE dispersions. Centrifugation speed serves as the primary selection parameter, enabling the classification of different production techniques into three aspect ratio categories. Moreover, the optical properties of the nanosheets vary as a function of their thickness.

Finally, the last two sections explore how networks formed through printing or polymer composite fabrication exhibit transport mechanisms influenced by both the intrinsic properties of the nanosheets and the junctions between them. Additionally, the various production techniques are briefly reviewed, highlighting their advantages and limitations. This chapter concludes with a summary of the key applications of these networks in applied science.

Conflict of interest

The authors declare no conflict of interest.

Author details

Daniel Olaya-Cortes*† and Yenny Hernandez†
Nanomaterials Laboratory, Physics Department, Universidad de los Andes, Bogota, Colombia

*Address all correspondence to: de.olaya1318@uniandes.edu.co

† These authors contributed equally.

References

[1] Dickinson RG, Pauling L. The crystal structure of molybdenite. Journal of the American Chemical Society. 1923;**45**(6): 1466-1471. DOI: 10.1021/ja01659a020

[2] Splendiani A, Sun L, Zhang Y, Li T, Kim J, Chim CY, et al. Emerging photoluminescence in monolayer MoS2. Nano Letters. 2010;**10**(4):1271-510.1021/nl903868w

[3] Zhan Y, Liu Z, Najmaei S, Ajayan PM, Lou J. Large-area vapor-phase growth and characterization of MoS2 atomic layers on a SiO2 substrate. Small. 2012;**8** (7):966-971. Available from: https://onlinelibrary.wiley.com/doi/abs/10.1002/smll.201102654

[4] Coleman JN, Lotya M, O'Neill A, Bergin SD, King PJ, Khan U, et al. Two-dimensional nanosheets produced by liquid exfoliation of layered materials. Science. 2011;**331**(6017):568-571. Available from: https://www.science.org/doi/10.1126/science.1194975

[5] Tenne R, Margulis L, Genut M, Hodes G. Polyhedral and cylindrical structures of tungsten disulphide. Nature. 1992;**360** (6403):444-446. Available from: https://www.nature.com/articles/360444a0

[6] You X, Liu N, Lee CJ, Pak JJ. An electrochemical route to MoS2 nanosheets for device applications. Materials Letters. 2014;**121**:31-35. Available from: https://www.sciencedirect.com/science/article/pii/S0167577X14000688

[7] El Garah M, Bertolazzi S, Ippolito S, Eredia M, Janica I, Melinte G, et al. MoS2 nanosheets via electrochemical lithium-ion intercalation under ambient conditions. FlatChem. 2018;**9**:33-39. Available from: https://www.sciencedirect.com/science/article/pii/S245226271830031X

[8] Olaya-Cortes D, Ramos L, Rubio T, Pabon JP, Hernandez Y. Lateral size dependence of photoconductivity in TMD networks. Journal of Physics: Condensed Matter. 2024;**37**(7):075301. DOI: 10.1088/1361-648X/ad965a

[9] Olaya-Cortes D, Navarrete C, Rubio T, Giraldo-Gallo P, Hernandez Y. Transition of piezoresistive gauge factor tuned by lateral size in MoS2 and WS2 networks. 2D Materials. 2025;**12**(2): 025002. DOI: 10.1088/2053-1583/ada2fb

[10] Yuan H, Liu X, Ma L, Gong P, Yang Z, Wang H, et al. High efficiency shear exfoliation for producing high-quality, few-layered MoS2 nanosheets in a green ethanol/water system. RSC Advances. 2016;**6**(86):82763-82773. Available from: https://pubs.rsc.org/en/content/articlelanding/2016/ra/c6ra15310k

[11] Cunningham G, Lotya M, Cucinotta CS, Sanvito S, Bergin SD, Menzel R, et al. Solvent exfoliation of transition metal dichalcogenides: Dispersibility of exfoliated nanosheets varies only weakly between compounds. ACS Nano. 2012;**6** (4):3468-3480. DOI: 10.1021/nn300503e

[12] Backes C, Szyd owska BM, Harvey A, Yuan S, Vega-Mayoral V, Davies BR, et al. Production of highly monolayer enriched dispersions of liquid-exfoliated nanosheets by liquid cascade centrifugation. ACS Nano. 2016;**10**(1): 1589-1601. DOI: 10.1021/acsnano.5b07228

[13] Kang J, Seo JWT, Alducin D, Ponce A, Yacaman MJ, Hersam MC. Thickness sorting of two-dimensional transition metal dichalcogenides via copolymer-assisted density gradient ultracentrifugation. Nature Communications. 2014;**5**(1):5478.

Available from: https://www.nature.com/articles/ncomms6478

[14] Kelly AG, Hallam T, Backes C, Harvey A, Esmaeily AS, Godwin I, et al. All-printed thin-film transistors from networks of liquid-exfoliated nanosheets. Science. 2017;**356**(6333):69-73. Available from: https://www.science.org/doi/10.1126/science.aal4062

[15] Biccai S, Boland CS, O'Driscoll DP, Harvey A, Gabbett C, O'Suilleabhain DR, et al. Negative gauge factor piezoresistive composites based on polymers filled with MoS2 nanosheets. ACS Nano. 2019;**13**(6):6845-6855. DOI: 10.1021/acsnano.9b01613

[16] Gabbett C, Kelly AG, Coleman E, Doolan L, Carey T, Synnatschke K, et al. Understanding how junction resistances impact the conduction mechanism in nano-networks. Nature Communications. 2024;**15**(1):4517. Available from: https://www.nature.com/articles/s41467-024-48614-5

[17] Zribi R, Foti A, Donato MG, Gucciardi PG, Neri G. Electrochemical and sensing properties of 2D-MoS2 nanosheets produced via liquid cascade centrifugation. Electrochimica Acta. 2022;**436**:141433. Available from: https://www.sciencedirect.com/science/article/pii/S0013468622015900

[18] Zhuo Y, Kinloch IA, Bissett MA. Simultaneous electrochemical exfoliation and functionalization of 2H-MoS2 for supercapacitor electrodes. ACS Applied Nano Materials. 2023;**6**(19):18062-18070. DOI: 10.1021/acsanm.3c03322

[19] Adilbekova B, Lin Y, Yengel E, Faber H, Harrison G, Firdaus Y, et al. Liquid phase exfoliation of MoS2 and WS2 in aqueous ammonia and their application in highly efficient organic solar cells. Journal of Materials Chemistry C. 2020;**8**(15):5259-5264. Available from: https://pubs.rsc.org/en/content/articlelanding/2020/tc/d0tc00659a

[20] Lin Z, Liu Y, Halim U, Ding M, Liu Y, Wang Y, et al. Solution-processable 2D semiconductors for high-performance large-area electronics. Nature. 2018;**562**(7726):254-258. Available from: https://www.nature.com/articles/s41586-018-0574-4

[21] Woodward RI, Kelleher EJR, Howe RCT, Hu G, Torrisi F, Hasan T, et al. Tunable Q-switched fiber laser based on saturable edge-state absorption in few-layer molybdenum disulfide (MoS2). Optics Express. 2014;**22**(25):31113-31122. Available from: https://opg.optica.org/oe/abstract.cfm?uri=oe-22-25-31113

[22] Novoselov KS, Geim AK, Morozov SV, Jiang D, Zhang Y, Dubonos SV, et al. Electric field effect in atomically thin carbon films. Science. 2004;**306**(5696):666-669. Available from: https://www.science.org/doi/10.1126/science.1102896

[23] Hernandez Y, Nicolosi V, Lotya M, Blighe FM, Sun Z, De S, et al. High-yield production of graphene by liquid-phase exfoliation of graphite. Nature Nanotechnology. 2008;**3**(9):563-568. Available from: https://www.nature.com/articles/nnano.2008.215

[24] McManus D, Santo AD, Selvasundaram PB, Krupke R, LiBassi A, Casiraghi C. Photocurrent study of all-printed photodetectors on paper made of different transition metal dichalcogenide nanosheets. Flexible and Printed Electronics. 2018;**3**(3):034005. DOI: 10.1088/2058-8585/aaddb5

[25] Liu Y, Li R. Study on ultrasound-assisted liquid-phase exfoliation for preparing graphene-like molybdenum disulfide nanosheets. Ultrasonics

Sonochemistry. 2020;**63**:104923. Available from: https://www.sciencedirect.com/science/article/pii/S1350417719314609

[26] Philip A, Kumar AR. Solvent effects on the electrochemical performance of few layered MoS2 electrodes fabricated using FTO substrates. Nano Express. 2024;**5**(1):015015. DOI: 10.1088/2632-959X/ad2b82

[27] Neilson J, Avery MP, Derby B. Tiled monolayer films of 2D molybdenum disulfide nanoflakes assembled at liquid/liquid interfaces. ACS Applied Materials and Interfaces. 2020;**12**(22):25125-25134. DOI: 10.1021/acsami.0c03794

[28] Mohammadpour Z, Abdollahi SH, Safavi A. Sugar-based natural deep eutectic mixtures as green intercalating solvents for high-yield preparation of stable MoS2 nanosheets: Application to electrocatalysis of hydrogen evolution reaction. ACS Applied Energy Materials. 2018;**1**(11):5896-5906. DOI: 10.1021/acsaem.8b00838

[29] Arshad MU, Raza H, Khan MB, Hussain A. Synthesis of 2D molybdenum disulfide (MoS2) for enhancement of mechanical and electrical properties of polystyrene (PS) polymer. Polymer Testing. 2020;**90**:106646. Available from: https://www.sciencedirect.com/science/article/pii/S0142941820304219

[30] Backes C, Higgins TM, Kelly A, Boland C, Harvey A, Hanlon D, et al. Guidelines for exfoliation, characterization and processing of layered materials produced by liquid exfoliation. Chemistry of Materials. 2017;**29**(1):243-255. DOI: 10.1021/acs.chemmater.6b03335

[31] Wang K, Wang J, Fan J, Lotya M, O'Neill A, Fox D, et al. Ultrafast saturable absorption of two-dimensional MoS2 nanosheets. ACS Nano. 2013;7(10):9260-9267. DOI: 10.1021/nn403886t

[32] Cunningham G, Khan U, Backes C, Hanlon D, McCloskey D, Donegan JF, et al. Photoconductivity of solution-processed MoS2 films. Journal of Materials Chemistry C. 2013;**1**(41):6899-6904. Available from: https://pubs.rsc.org/en/content/articlelanding/2013/tc/c3tc31402b

[33] Finn DJ, Lotya M, Cunningham G, Smith RJ, McCloskey D, Donegan JF, et al. Inkjet deposition of liquid-exfoliated graphene and MoS2 nanosheets for printed device applications. Journal of Materials Chemistry C. 2014;**2**(5):925-932. Available from: https://pubs.rsc.org/en/content/articlelanding/2014/tc/c3tc31993h

[34] Kaushik V, Wu S, Jang H, Kang J, Kim K, Suk JW. Scalable exfoliation of bulk MoS2 to single- and few-layers using Toroidal Taylor vortices. Nanomaterials. 2018;**8**(8):587. Available from: https://www.mdpi.com/2079-4991/8/8/587

[35] Liu N, Kim P, Kim JH, Ye JH, Kim S, Lee CJ. Large-area atomically thin MoS2 nanosheets prepared using electrochemical exfoliation. ACS Nano. 2014;**8**(7):6902-6910. DOI: 10.1021/nn5016242

[36] Tan X, Kang W, Liu J, Zhang C. Synergistic exfoliation of MoS2 by ultrasound sonication in a supercritical fluid based complex solvent. Nanoscale Research Letters. 2019;**14**(1):317. DOI: 10.1186/s11671-019-3126-4

[37] Zeng X, Hirwa H, Metel S, Nicolosi V, Wagner V. Solution processed thin film transistor from liquid phase exfoliated MoS2 flakes. Solid-State Electronics. 2018;**141**:58-64. Available

from: https://www.sciencedirect.com/science/article/pii/S0038110117304185

[38] Li J, Naiini MM, Vaziri S, Lemme MC, Östling M. Inkjet printing of MoS2. Advanced Functional Materials. 2014;**24**(41):6524-6531. Available from: https://onlinelibrary.wiley.com/doi/abs/10.1002/adfm.201400984

[39] Martínez-Jódar A, Villar-Rodil S, Munuera JM, Castro-Muñiz A, Coleman JN, Raymundo-Piñero E, et al. Two-dimensional MoS2 nanosheets derived from cathodic exfoliation for lithium storage applications. Nanomaterials. 2024;**14**(11):932. Available from: https://www.mdpi.com/2079-4991/14/11/932

[40] Shang J, Xue F, Ding E. Efficient exfoliation of molybdenum disulphide nanosheets by a high-pressure homogeniser. Micro and Nano Letters. 2015;**10**(10):589-591. Available from: https://onlinelibrary.wiley.com/doi/abs/10.1049/mnl.2015.0173

[41] Wang Y, Wang K, Zhang C, Zhu J, Xu J, Liu T. Solvent-exchange strategy toward aqueous dispersible MoS2 nanosheets and their nitrogen-rich carbon sphere nanocomposites for efficient lithium/sodium ion storage. Small. 2019;**15**(45):1903816. Available from: https://onlinelibrary.wiley.com/doi/abs/10.1002/smll.201903816

[42] Alzakia FI, Jonhson W, Ding J, Tan SC. Ultrafast exfoliation of 2D materials by solvent activation and one-step fabrication of all-2D-material Photodetectors by Electrohydrodynamic printing. ACS Applied Materials and Interfaces. 2020;**12**(25):28840-28851. DOI: 10.1021/acsami.0c06279

[43] Muscuso L, Cravanzola S, Cesano F, Scarano D, Zecchina A. Optical, vibrational, and structural properties of MoS2 nanoparticles obtained by exfoliation and fragmentation via ultrasound cavitation in isopropyl alcohol. The Journal of Physical Chemistry C. 2015;**119**(7):3791-3801. DOI: 10.1021/jp511973k

[44] Howe RCT, Woodward RI, Hu G, Yang Z, Kelleher EJR, Hasan T. Surfactant-aided exfoliation of molybdenum disulfide for ultrafast pulse generation through edge-state saturable absorption. Physica Status Solidi (b). 2016;**253**(5):911-917. Available from: https://onlinelibrary.wiley.com/doi/abs/10.1002/pssb.201552304

[45] Sim DM, Han HJ, Yim S, Choi MJ, Jeon J, Jung YS. Long-term stable 2H-MoS2 dispersion: Critical role of solvent for simultaneous phase restoration and surface functionalization of liquid-exfoliated MoS2. ACS Omega. 2017;**2**(8):4678-4687. DOI: 10.1021/acsomega.7b00841

[46] Zhou L, He B, Yang Y, He Y. Facile approach to surface functionalized MoS2 nanosheets. RSC Advances. 2014;**4**(61):32570-32578. Available from: https://pubs.rsc.org/en/content/articlelanding/2014/ra/c4ra04682j

[47] Ambrosi A, Pumera M. Electrochemical exfoliation of MoS2 crystal for hydrogen electrogeneration. Chemistry – A European Journal. 2018;**24**(69):18551-18555. Available from: https://onlinelibrary.wiley.com/doi/abs/10.1002/chem.201804821

[48] Kajbafvala M, Farbod M. Effective size selection of MoS2 nanosheets by a novel liquid cascade centrifugation: Influences of the flakes dimensions on electrochemical and photoelectrochemical applications. Journal of Colloid and Interface Science. 2018;**527**:159-171. Available from: https://www.sciencedirect.com/science/article/pii/S0021979718305423

[49] Backes C, Smith RJ, McEvoy N, Berner NC, McCloskey D, Nerl HC, et al. Edge and confinement effects allow in situ measurement of size and thickness of liquid-exfoliated nanosheets. Nature Communications. 2014;**5**(1):4576. Available from: https://www.nature.com/articles/ncomms5576

[50] Ogilvie SP, Large MJ, O'Mara MA, Lynch PJ, Lee CL, King AAK, et al. Size selection of liquid-exfoliated 2D nanosheets. 2D Materials. 2019;**6**(3): 031002. DOI: 10.1088/2053-1583/ab0dc3

[51] Kelly AG, O'Suilleabhain D, Gabbett C, Coleman JN. The electrical conductivity of solution-processed nanosheet networks. Nature Reviews Materials. 2022;**7**(3):217-234. Available from: https://www.nature.com/articles/s41578-021-00386-w

[52] Eda G, Chhowalla M. Graphene-based composite thin films for electronics. Nano Letters. 2009;**9**(2): 814-818. DOI: 10.1021/nl8035367

[53] Kelly AG, Finn D, Harvey A, Hallam T, Coleman JN. All-printed capacitors from graphene-BN-graphene nanosheet heterostructures. Applied Physics Letters. 2016;**109**(2):023107. DOI: 10.1063/1.4958858

[54] Cunningham G, Hanlon D, McEvoy N, Duesberg GS, Coleman JN. Large variations in both dark- and photoconductivity in nanosheet networks as nanomaterial is varied from MoS2 to WTe2. Nanoscale. 2014;**7**(1): 198-208. Available from: https://pubs.rsc.org/en/content/articlelanding/2015/nr/c4nr04951a

[55] Pavličková M, Lorencová L, Hatala M, Kováč M, Tkáč J, Gemeiner P. Facile fabrication of screen-printed MoS2 electrodes for electrochemical sensing of dopamine. Scientific Reports. 2022;**12**(1):11900. Available from: https://www.nature.com/articles/s41598-022-16187-2

[56] Yakimchuk E, Soots R, Kotin I, Antonova I. 2D printed graphene conductive layers with high carrier mobility. Current Applied Physics. 2017; **17**(12):1655-1661. Available from: https://www.sciencedirect.com/science/article/pii/S1567173917302638

[57] Aigner W, Bienek O, Falcão BP, Ahmed SU, Wiggers H, Stutzmann M, et al. Intra- and inter-nanocrystal charge transport in nanocrystal films. Nanoscale. 2018;**10**(17):8042-8057. Available from: https://pubs.rsc.org/en/content/articlelanding/2018/nr/c8nr00250a

Chapter 6

Nonlinear Optical Effect and Its Applications of TMDs

Xiangling Tian, Lun Wang and Rongfei Wei

Abstract

This chapter examines the nonlinear optical properties and applications of transition metal dichalcogenides (TMDs), underscoring their importance in advanced photonic and optoelectronic devices. The discussion begins with an introduction to the fundamental mechanisms underlying second- and third-order NLO processes, including second harmonic generation, third harmonic generation, and two-photon absorption. Furthermore, the chapter provides a comprehensive overview of experimental techniques, such as the Z-scan method, pump-probe spectroscopy, and multiphoton microscopy imaging, utilized to characterize these phenomena effectively. The distinctive characteristics of TMDs, including their ultra-broadband absorption and substantial NLO responses, are highlighted, with a particular emphasis on their applications in optical limiting, saturable absorption, pulsed laser generation, and biological sensors. In addition, the integration of TMDs into emerging technologies, such as optical modulators, THz radiation generation, and nonlinear holography, is explored. The chapter concludes by discussing their biological applications, including two-photon luminescence imaging and surface-enhanced Raman spectroscopy, thereby demonstrating the broad versatility of TMDs within various scientific and technological contexts.

Keywords: nonlinear optical, TMDs, second order NLO, third order NLO, NLO application

1. Introduction

Optics, which originated in the eighteenth century, initially focused on natural light sources. The advent of the laser in 1960 marked a pivotal transition to modern optics [1], characterized by the utilization of stimulated emission light sources. Lasers produce photons that are identical in state, thereby facilitating the development of the field of photonics. Photonics deals with the generation, transmission, manipulation, detection, display, storage, and interaction of photons with matter, serving as a fundamental pillar of contemporary optics.

A notable advancement within this domain is "Nanophotonics," which integrates nanotechnology with ultrafast photonic techniques. The primary objective of nanophotonics is to achieve the concept of "controlling light with light," often referred to as all-optical technology. However, photons are devoid of charge and cannot interact directly in the same manner as electrons within a transistor. To address this limitation, researchers employ nonlinear optical methods. These techniques utilize pump light to modify the properties or microstructure of a medium, thereby regulating the frequency, amplitude, phase, polarization, or group velocity of signal light as it propagates through the medium [2, 3].

The field of nonlinear optics is often considered to have commenced with the discovery of second-harmonic generation (SHG) by Franken *et al.* in 1961 [4], following Maiman's demonstration of the first operational laser in 1960. Nonlinear optics (NLO) investigates phenomena that occur when the optical properties of a material system are altered by intense light. Low-intensity light sources, such as sunlight or fluorescent lamps, do not elicit nonlinear optical responses and instead exhibit linear behavior as governed by superposition principles. In contrast, focused laser beams can achieve intensities of $10^{15}$to 10^{22} W/cm^2, generating electric fields comparable to interatomic electric fields (10^5 to 10^8 V/m), thus satisfying the conditions requisite for nonlinear optical effects [5].

Since the successful isolation of atomic-layer-thick graphene *via* mechanical exfoliation, materials such as MoS_2 and its analogs have garnered increasing attention from the research community. The atomically thin nature of 2D materials imparts unique electronic and optical properties, attributable to reduced dielectric shielding and enhanced Coulomb interactions [6]. Such materials exhibit advantages, including ultra-broadband absorption and pronounced nonlinear optical responses, making them highly suitable for various nonlinear applications. These applications encompass optical communication, frequency conversion, imaging, electro-optic modulation, mode-locking, and Q-switching. The exploration and enhancement of the nonlinear optical properties of MoS_2-based materials are pivotal in advancing optical and optoelectronic devices.

In this chapter, we will first present a comprehensive overview of the fundamental physical mechanisms underlying second- and third-order nonlinear optical processes. Subsequently, we will focus on methodologies for characterizing the nonlinear optical properties of MoS_2-based materials. Finally, we will explore several significant applications that leverage the NLO properties of these materials.

2. Fundamental physical mechanisms

2.1 Basics of NLO

In a nonlinear material system, the response to an incident optical field is not directly proportional to the light intensity. Instead, it exhibits a nonlinear dependence. This nonlinear behavior can be described macroscopically by expanding the polarization (t) in terms of the incident electric field strength $E(t)$ as a power series. The series can be written as:

$$P(t) = \varepsilon_0 \left(\chi^{(1)} E(t) + \chi^{(2)} E(t) + \chi^{(3)} E(t) + \dots \right) \quad (1)$$

In this expression, $\chi^{(n)}$ denotes the n-th order nonlinear susceptibility. The first term, $\chi^{(1)}$, accounts for linear optical effects, such as refraction and absorption, where the polarization of the material is directly proportional to the electric field strength, given that the field intensity $E(t)$ is relatively weak. However, when (t) becomes sufficiently strong, or if the nonlinear susceptibility is large, higher-order terms begin to have a significant impact. The second term, $\chi^{(2)}$, corresponds to the second-order nonlinear susceptibility and governs second-order nonlinear optical processes. The third term, involving $\chi^{(3)}$, describes third-order nonlinearities. For $n\geq4$, more complex phenomena like high harmonic generation (HHG) can emerge, though this chapter does not focus on HHG in detail.

Nobel laureate Bloembergen [7], a leading expert in nonlinear optics, provided a precise scientific definition of the field: "If the response of matter to the impressed electromagnetic field is not a linear function of the impressed electromagnetic field strength, this optical phenomenon belongs to the field of nonlinear optics".

It is important to note that the definition refers to the electromagnetic field, which consists of both electric and magnetic fields. While the magnetic component of light does influence the magnetic polarization of a material, its effect on TMDs is negligible and can be ignored. Therefore, we focus solely on the contribution of the electric field to the nonlinear optical effects in the medium.

Considering only the linear and third-order nonlinear optical effects, the total polarization of the medium is:

$$P = \varepsilon_0\left(\chi^{(1)} + \chi^{(3)}|E|^2\right)E = \varepsilon_0\left(\chi^{(1)} + \Delta\chi^{(1)}\right)E \tag{2}$$

where

$$\Delta\chi^{(1)} = \chi^{(3)}|E|^2 \propto \chi^{(3)} I \tag{3}$$

In this context, $\Delta^{(1)}$ represents the change in first-order susceptibility induced by the electric field of the incident light. This variation is proportional to the square of the light field amplitude, or equivalently, the light intensity I. Since the third-order nonlinear susceptibility is a complex quantity, it can be expressed as having both real and imaginary components:

$$\chi^{(3)} = \chi^{(3)\prime} + i\chi^{(3)\prime\prime} \tag{4}$$

The refractive index and absorption coefficient of a medium can be separated into linear and nonlinear components:

$$n = n_0 + \Delta n \tag{5}$$

$$\alpha = \alpha_0 + \Delta\alpha \tag{6}$$

where n_0 represents the linear refractive index, Δn is the nonlinear refractive index, α_0 is the linear absorption coefficient, and $\Delta\alpha$ is the nonlinear absorption coefficient. The nonlinear refractive index is related to the real part of the susceptibility, while the nonlinear absorption coefficient is connected to the imaginary part of the susceptibility. Both of these quantities are proportional to light intensity $I \propto |E|^2$:

$$\Delta n \propto \left(\chi^{(3)\prime}\right) I \tag{7}$$

$$\Delta\alpha = \left(\chi^{(3)\prime\prime}\right) I \tag{8}$$

In summary, for third-order nonlinear effects, the changes in susceptibility, refractive index, and absorption coefficient are all linearly dependent on light intensity, or more specifically, on the optical field amplitude. Therefore, NLO phenomena can generally be described as the variation of optical parameters in response to changes in the optical field amplitude.

2.2 Typical NLO

Nonlinear optical processes in TMDs can be broadly categorized as parametric and nonparametric. Parametric processes involve the excitation of ground states to virtual states, where the initial and final quantum states remain identical, with no absorption or electron transfer between virtual states. In contrast, nonparametric processes involve excitation to real states, where absorption and electron transfer occur. Parametric processes are described by real susceptibility, while nonparametric processes require complex susceptibility for their characterization.

In TMDs, parametric processes primarily include scattering phenomena such as SHG, sum-frequency generation (SFG), difference-frequency generation (DFG), third harmonic generation (THG), four-wave mixing (FWM), and stimulated Raman scattering [8–10]. Nonparametric processes, on the other hand, are characterized by effects such as saturable absorption (SA), two-photon absorption (TPA), self-focusing, self-defocusing, optical limiting, and the optical Kerr effect [8–11]. This section provides an overview of these nonlinear processes.

2.2.1 Second Oder phenomena

In Eq. (1), $\chi^{(2)}$ represents the second-order nonlinear susceptibility, which governs second-order nonlinear optical processes. These processes are critical for nonlinear photonic applications, enabling optical frequency conversion, the generation of new wavelengths, and the amplification of weak optical signals.

2.2.1.1 SHG

SHG is a type of wave-mixing process where an optical wave interacts with itself, producing a new wave at twice the original frequency. This means that part of the energy from a light wave at frequency ω is converted into a wave at 2ω, as illustrated in **Figure 1a**.

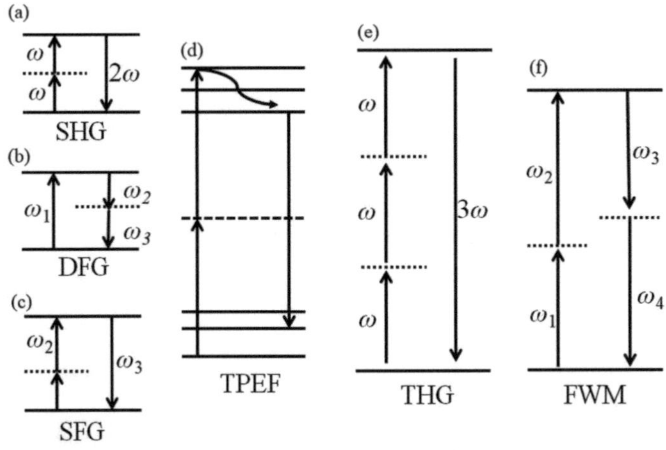

Figure 1.
Schematics of SHG, DFG, SFG, TPEF, and FWM.

To qualitatively describe SHG, when a laser beam with electric field strength $E(t) = Ee^{-i\omega t} + c.c.$ interacts with a nonlinear material, the nonlinear polarization includes two components:

$$P^2(t) = 2\varepsilon_0\chi^2 EE^* + \left(\varepsilon_0\chi^2 E^2 e^{-i2\omega t} + c.c.\right) \quad (9)$$

one at zero frequency and another at 2ω. The 2ω component contributes to the generation of the second harmonic wave. This process also results in optical rectification, creating a static electric field across the nonlinear crystal. Under optimal conditions, such as with a Nd:YAG laser, SHG can efficiently convert all incident light at ω into 2ω. This phenomenon can also be understood in terms of photon exchange among various frequency components.

2.2.1.2 DFG and SFG

Three-wave mixing is a nonlinear optical process where two incident waves with different frequencies (ω_1 and ω_2, where $\omega_1 \neq \omega_2$) interact in a nonlinear medium to generate a third wave with a new frequency (ω_3). This process is categorized as either DFG or SFG based on the relationship between the input and output frequencies, as illustrated in **Figure 1b** and **c**.

DFG occurs when the generated frequency is the difference between the two incident frequencies, given by $\omega_3 = \omega_1 - \omega_2$.

SFG, on the other hand, arises when the output frequency equals the sum of the two incident frequencies, $\omega_3 = \omega_1 + \omega_2$.

The nonlinear (NL) polarization complex amplitude can be expressed as follows:

$$P(\omega_1 + \omega_2) = 2\varepsilon_0\chi^2 E_1 E_2, P(\omega_1 - \omega_2) = 2\varepsilon_0\chi^2 EE^* \quad (10)$$

In this expression, the applied and generated frequencies are distinct. DFG is commonly used to produce tunable infrared radiation by combining a visible-region frequency with a fixed-frequency laser. While DFG and SFG share similarities, a key distinction lies in their description using photon energy levels, where the conservation of energy must always hold.

DFG is also referred to as optical parametric amplification. During the process, a photon of frequency ω_1 is absorbed by an atom, exciting it to a higher virtual energy state. From there, stimulated emission occurs. When an external field at ω_2 is applied, two-photon emission (TPE) can be facilitated. However, if this external field is absent, the resulting emission is much weaker, as it originates from spontaneous TPE.

2.2.1.3 Two-photon excitation fluorescence (TPEF)

TPEF microscopy was first introduced by Denk et al. in 1990 [12]. In this process, when the combined energy of two incident photons matches the transition frequency of a target molecule, the molecule absorbs both photons simultaneously, reaching an excited electronic state. It then emits a single fluorescent photon with higher energy than either of the individual incident photons. This phenomenon is known as TPEF, and the energy diagram is illustrated in **Figure 1d**.

TPEF microscopy is a powerful fluorescence imaging technique. Unlike conventional fluorescence microscopy, where the emission wavelength is longer than the excitation wavelength, the emitted light in TPEF has a shorter wavelength than the combined excitation photons. Additionally, TPEF has a small excitation volume, which significantly reduces photodamage to the sample and minimizes photobleaching of the fluorophore. The use of longer excitation wavelengths in TPEF enhances tissue penetration, reaching depths of up to approximately 1 mm, while also reducing photolysis-related damage.

A key advantage of TPEF is its quadratic dependence on the excitation light intensity, which enables precise optical sectioning. Furthermore, due to the multiphoton absorption process, background signals are strongly suppressed. It is worth noting that the TPEF emission wavelength is longer than that of SHG. Experimental results indicate that TPEF and SHG images exhibit comparable signal quality.

In addition to DFG, SFG and TPEF, other significant second-order nonlinear optical processes include optical rectification, the Pockels effect, optical parametric amplification, and optical parametric oscillation. These phenomena have diverse applications in photonics and are key to understanding the broader implications of second-order nonlinearities.

2.2.2 Third Oder phenomena

The $\chi^{(3)}$ term in Eq. (1) represents the third-order nonlinear susceptibility, which gives rise to various nonlinear optical (NLO) phenomena. These include THG, FWM, TPA, self-phase modulation, the optical Kerr effect, and stimulated Raman scattering (SRS).

2.2.2.1 THG

THG is a type of third-order nonlinear effect. Similar to SHG, it involves frequency conversion, but in this case, three photons with frequency ω interact within a medium to produce a single photon with frequency 3ω. This process is expressed mathematically through the nonlinear polarization, as shown in Eq. (11), and illustrated in **Figure 1e**.

$$P^3(t) = \frac{1}{4}\varepsilon_0\chi^3\varepsilon^3 \cos 3\omega t + \frac{3}{4}\varepsilon_0\chi^3\varepsilon^3 \cos\omega t \tag{11}$$

In this process, three photons at ω are annihilated, and one photon at 3ω is created. Unlike SHG, THG does not require the material to lack inversion symmetry, although it does depend on the medium's third-order susceptibility and requires finite phase-matching in the excitation region. As a result, THG can occur in both centrosymmetric and non-centrosymmetric materials. While its functionality is similar to SHG, THG exhibits a cubic dependence on the intensity of the excitation light.

2.2.2.2 FWM

FWM is a third-order NLO process in which photons of two or three different wavelengths interact within a material to generate new photons at different wavelengths. This process is independent of the material's inversion symmetry, meaning FWM can theoretically occur in any medium.

FWM can take on various forms, one example of which is illustrated in **Figure 1f**. In this scenario, the generated photon frequencies satisfy the relationship $\omega_4=\omega_1+\omega_2-\omega_3$. The complex amplitude of the nonlinear polarization involved in FWM can be described by Eq. (12).

$$P(\omega_1 + \omega_2 - \omega_3) = 6\varepsilon_0 \chi^3 E_1 E_2 E_3^* \tag{12}$$

This versatile phenomenon has a wide range of applications, including optical imaging, wavelength conversion, and other advanced optical processing techniques.

2.2.2.3 Kerr effect, self-focusing and self-defocusing

The Kerr effect is a form of elastic nonlinearity, meaning no energy is transferred from light to the material during the process. It arises from the anharmonic motion of electrons in a strong electromagnetic field. Under an applied electric field E, the total polarization P can be described by Eq. (1), which accounts for both linear and nonlinear responses. The real part of $\chi^{(3)}$ represents the Kerr effect.

In fiber optics, the refractive index n influenced by the Kerr effect can be expressed as [13]:

$$n = n_0 + n_2 I \tag{13}$$

where n_0 is the linear refractive index, n_2 is the nonlinear refractive index coefficient, and I is the optical intensity. The sign of n_2, which depends on the medium, determines whether the refractive index increases or decreases with higher light intensity. In contrast, for linear optics, the refractive index remains constant at n_0, regardless of light intensity.

For layered materials, the real part of $\chi^{(3)}$ is expressed as:

$$Re\,\chi^{(3)}_{monolayer} = \frac{n_0^2 n_2 (cm^2/W)}{0.0395 N_{eff}^2} \tag{14}$$

where $\chi^{(3)}_{monolayer}$ is the susceptibility of a single monolayer and N_{eff} is the effective number of layers. The Kerr effect is represented by the total real part $\chi^{(3)}_{total} = N_{eff}^2 \chi^{(3)}_{monolayer}$.

When a Gaussian laser beam passes through a medium, the radial variation in light intensity causes a corresponding change in the refractive index due to the Kerr effect. For $n_2 > 0$, this creates a lens-like focusing effect (self-focusing), reducing the beam's transverse dimension. Conversely, for $n_2 < 0$, a lens-like defocusing effect (self-defocusing) occurs, causing the beam to expand. These nonlinear refraction effects are illustrated in **Figure 2**.

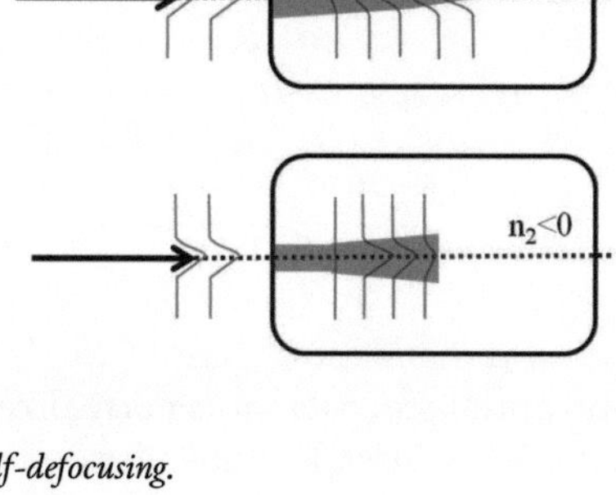

Figure 2.
Illustration of self-focusing and self-defocusing.

2.2.2.4 SA

The imaginary part of $\chi^{(3)}$ represents nonlinear absorption, an inelastic process where energy is exchanged between light and the material. Nonlinear absorption describes how the absorption coefficient, α, which is a parameter in Beer's law, varies with the intensity of light. In its simplest form, nonlinear absorption can be expressed as a power series:

$$\frac{dI}{dz} = -\alpha(I)I \tag{15}$$

(where) $\alpha(I) = \alpha + \alpha^{(2)}I + \alpha^{(3)}I^2 + \cdots$

Here, $\alpha^{(2)}$ is the two-photon absorption coefficient (commonly reported in units like cm GW^{-1}), while $\alpha^{(3)}$ is the three-photon absorption coefficient, and so on. This equation is also used in cases of sequential absorption, such as reverse saturable absorption (RSA). For SA, $\alpha^{(2)}$ is often assigned a negative value to reflect the decrease in absorption.

In two-dimensional materials, SA arises from a phenomenon called Pauli blocking. When electrons are excited by light into the conduction band, scattering may cause some energy loss. As the conduction band fills faster than electrons can relax, the material becomes momentarily "saturated" or "bleached." The absorption coefficient can then be expressed as:

$$\alpha = \alpha_{ns} + \frac{\alpha_s}{1 + I/I_{sat}} \tag{16}$$

where α_{ns} and α_s are the absorption coefficients before and after saturation, I is the incident intensity, and I_{sat} is the saturation intensity.

Processes like SA, RSA, and excited state absorption (ESA) often overlap with multiphoton absorption [14]. While these effects may not strictly depend on the nonlinear polarization response to the electric field, their intensity- or fluence-dependent behavior warrants their inclusion in discussions of nonlinear optics.

TPA is a specific nonlinear process where one electron simultaneously absorbs two photons (discussed in detail later) [14]. At high intensities, TPA becomes significant, and SA can be modeled as:

$$\alpha = \alpha_{ns} + \frac{\alpha_s}{1 + I/I_{sat}} + \beta I \tag{17}$$

where β is the TPA coefficient. The corresponding light transmission, T, is expressed as:

$$T = \exp(-\alpha) = exp\left(-\alpha_{ns} - \frac{\alpha_s}{1 + I/I_{sat}} - \beta I\right)$$
$$\approx exp\left(-\alpha_{ns}\right)\left(1 - \frac{\alpha_s}{1 + I/I_{sat}} - \beta I\right) = T_0 - \frac{T_0\alpha_s}{1 + I/I_{sat}} - T_0\beta I \tag{18}$$

Here, T_0 is the low-intensity transmission, defined as $T_0 = \exp.(-\alpha_{ns})$. T_o account for sample thickness, the absorption coefficient is normalized as $\alpha_0=\alpha/L$, where L is the thickness, and absorbance A is related to transmission by $T=10^{-A}$ and $A=0.434\alpha$.

However, the model described above often fails to match experimental results. This is because the term $(1+I/I_{sat})^{-1}$ requires a very broad intensity range, often exceeding a 100-fold increase, to effectively transition between unsaturated and saturated states. To address this limitation, a revised model is widely used:

$$T = T_0 - \Delta T exp(-I/I_{sat}) - T_0 \beta I \quad (19)$$

2.2.2.5 TPA

Comparing data from different sources can be challenging due to the dependence of nonlinear absorption parameters on laser pulse duration. For instance, some nonlinear processes are better described as fluence-dependent rather than intensity-dependent. Additionally, it is important to note that the absorption coefficient α, as well as higher-order terms like $\alpha^{(i)}$ (where i=2,3, ... i = 2,3, ...), are influenced by the material's concentration. Therefore, a more reliable method for evaluating the NLO properties of materials is to use microscopic parameters, such as absorption cross-sections, rather than macroscopic coefficients.

The principle of TPA is illustrated in **Figure 3a**. In this process, two photons are simultaneously absorbed, causing a transition from the ground state to an excited state. TPA is typically characterized by the two-photon absorption cross-section, σ_2, expressed in GM units (1 GM = 10^{-50} cm^4s) [15]. The relationship between the macroscopic two-photon coefficient and σ_2 is given by

$$\alpha^{(2)} = \frac{N\sigma_2}{\hbar\omega} \quad (20)$$

where N is the species concentration (in cm^{-3}).

In the TPA process, the transition rate R within an atom is proportional to the square of the laser intensity, as described by the relation:

$$R = \frac{\sigma I}{\hbar\omega} \quad (21)$$

This relationship makes TPA a valuable spectroscopic tool for determining the energy levels of materials relative to their ground state.

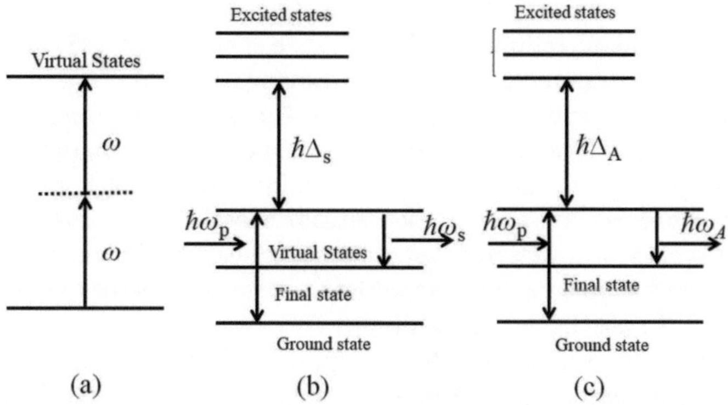

Figure 3.
The principle of TPA and Illustration of the SRS, the subscripts "S" and "A" stand for the Stokes and anti-Stokes modes, respectively.

2.2.2.6 SRS

In SRS, a photon with a frequency ω is absorbed, and a new photon with a lower frequency, called the Stokes frequency ($\omega_s=\omega-\omega_v$), is emitted. This process leaves the atoms in an excited state with energy $\hbar\omega_v$, which corresponds to the vibrational energy of the system. SRS was first observed in molecules, where it is closely tied to vibrational energy levels. Unlike spontaneous Raman scattering, SRS is highly efficient, achieving conversion efficiencies of over 10% of the incident light power. The Raman spectrum provides valuable information about a molecule's vibrational modes by measuring their density. In the spontaneous Raman process, photons from an incident laser interact with molecules in the sample and are scattered, producing both lower-frequency (Stokes) and higher-frequency (anti-Stokes) photons [10]. These two phenomena are illustrated in **Figure 3b** and **c**.

3. Experimental techniques

Characterizing the NLO properties of TMDs is essential, regardless of the methods used for designing and fabricating TMD-based active photonic devices. Over time, various experimental techniques have been developed to study the specific NLO properties of photonic materials. Both for advancing fundamental research and enabling practical applications, it is crucial to select appropriate methods for measuring the optical nonlinear susceptibility of these materials.

Several techniques are widely used to quantify optical nonlinearities, including *Z*-scan [16–18], multiphoton microscopy imaging [19–21], and pump-probe experiments [22–24]. These methods offer valuable insights into the distinct characteristics of NLO behavior in photonic materials. The choice of technique depends on the specific NLO parameters being investigated and the material's properties.

In the following sections, we provide a brief overview of some well-established methods for characterizing nonlinear optical properties.

3.1 *Z*-scan technique

The Z-scan method, introduced by Sheik-Bahae et al. in 1990 [25], is a widely used, straightforward, and accurate technique for studying the NLO properties of materials, including nonlinear absorption and refraction. It is particularly effective for analyzing third- and higher-order (odd) NLO processes, such as the imaginary part of $\chi^{(3)}$ (nonlinear absorption), the real part of $\chi^{(3)}$ (nonlinear refraction), and thermally induced optical nonlinearity. This method enables researchers to qualitatively generate a *Z*-scan curve, from which the sign of the nonlinear coefficient can be inferred based on the curve's shape. Further theoretical analysis allows for the quantitative determination of the coefficient's magnitude.

The *Z*-scan technique relies on detecting spatial distortions (both amplitude and phase) in the transmitted beam as it passes through the sample. In this setup, the sample is exposed to a beam with varying pump fluences, and the transmitted light is collected for analysis (**Figure 4a**). Two measurement modes are typically used: open aperture and closed aperture.

DOI: http://dx.doi.org/10.5772/intechopen.1009547

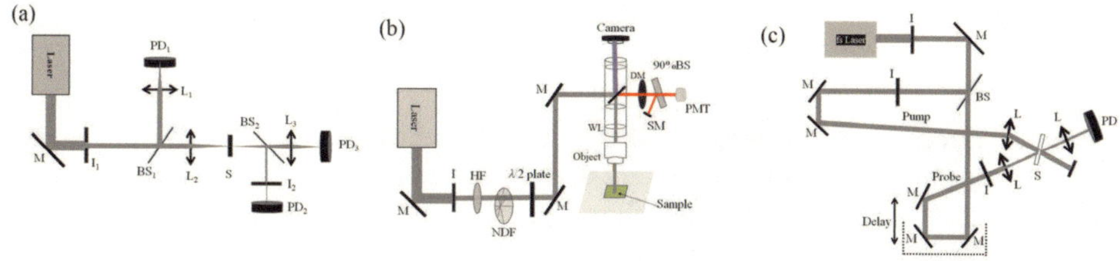

Figure 4.
Experimental techniques: (a) Z-scan experiment, (b) Pump-Probe setup, and (c) Multiphoton microscopy experiment.

3.1.1 Open aperture Z-scan

All transmitted light is collected using a convex lens (L3) to monitor intensity distortions of the transmitted beam (detected by PD3).

3.1.2 Closed aperture Z-scan

A small aperture (I2) is placed in front of the detector to capture phase distortions in the transmitted beam (detected by PD2).

Measurements are taken while moving the sample along the z-axis within the focal plane of a lens (L2). The nonlinear transmittance profile is obtained by comparing the signals from detectors (PD2 and PD3) to a reference photodiode (PD1). The collected data are then analyzed to quantitatively evaluate the sample's NLO properties.

Over time, the original *Z*-scan setup has been modified in various ways to suit different applications. Detailed discussions on these adaptations can be found in the work of De Araujo et al. [26].

To determine the real and imaginary components of a material's third-order nonlinear susceptibility, $\chi^{(3)}$, several assumptions are made: the beam is Gaussian, the sample is thin, the nonlinearity is instantaneous, phase changes are small, and the far-field condition is satisfied. Under these conditions, the relative axial transmittance of the sample can be expressed as [27, 28]:

$$T(z, \Delta\phi_0) \cong 1 - \frac{4\Delta\phi_0 z/z_0}{\left(z^2/z_0^2 + 9\right)\left(z^2/z_0^2 + 1\right)} \tag{22}$$

Here, z_0 is the Rayleigh range, z is the axial position, and $\Delta\phi_0$ represents the phase change, which is defined as [27, 28]:

$$\Delta\phi_0 = \frac{2\pi}{\lambda} L_{eff} n_2 I_0 \tag{23}$$

In this equation, I_0 is the peak intensity, λ is the wavelength of the radiation, and L_{eff} is the effective length of the sample, calculated as $L_{\text{eff}} = [1 - \exp.(-\alpha L)]/\alpha$, where α is the absorption coefficient.

The peak-to-valley transmittance variation can then be described as:

$$\Delta T_{p-v} = 0.406\Delta\phi_0 \tag{24}$$

The absorption coefficient β is determined using the following equation:

$$\beta = \frac{2\sqrt{2}\Delta T}{I_0 L_{eff}} \tag{25}$$

Here, ΔT represents the single valley value from the Z-scan curve. By combining the known sample and laser parameters with measurements of the far-field transmittance variation, the refractive index change can be calculated as [27–29]:

$$\mathrm{Re}\,\chi^{(3)}(\mathrm{esu}) = 10^{-4}\frac{\varepsilon_0 c^2 n_0^2 n_2}{\pi}\,(\mathrm{cm^2/W})$$

$$\mathrm{Im}\chi^{(3)}(\mathrm{esu}) = 10^{-2}\frac{\varepsilon_0 c^2 n_0^2 \lambda\beta}{4\pi^2}\,(\mathrm{cm^2/W})$$

$$\chi^{(3)}(esu) = \sqrt{\left(Re\,\chi^{(3)}(esu)\right)^2 + \left(Im\chi^{(3)}(esu)\right)^2} \tag{26}$$

The Z-scan method remains one of the most widely used techniques for characterizing the real and imaginary parts of $\chi^{(3)}$ in various materials.

3.2 Multiphoton microscopy imaging technique

Multiphoton microscopy imaging is a highly adaptable technique used to map samples with NLO responses, such as SHG, THG, or more wave interactions in nonlinear media that produce new frequencies.

The experimental setup for multiphoton microscopy is illustrated in **Figure 4b**. A pump beam of a specific wavelength is tightly focused onto the sample using a CCD camera and a microscope objective lens. The signal generated by the sample is separated using a series of wavelength-specific dichroic filters (LF, HF, and DM). To control the pump beam's intensity, a variable neutral density (ND) filter is employed, while a half-wave plate (HWP) is used to adjust the polarization of the beam.

In the schematic, a variable attenuator (VA) regulates the incident light power on the sample, and a quarter-wave plate (QWP) can transform linearly polarized light into circularly or elliptically polarized light. A 2D galvo scanner enables precise scanning of the laser beam, which is directed through a scan lens and a tube lens to the rear aperture of the objective lens.

The spectral characteristics and intensity of generated signals are recorded using ultraviolet (UV)-visible-near-infrared (NIR) spectrometers and a detector, such as a photomultiplier tube (PMT).

3.3 Pump-probe (PP) technique

Ultrafast pump-probe (PP) spectroscopy is a powerful technique for studying photo-induced processes occurring on timescales much shorter than the bandwidth of conventional electronic devices. It is widely applied to investigate ultrafast carrier dynamics, as well as charge and energy transfer processes. These mechanistic insights are valuable for optimizing the performance of ultrafast photonic devices.

The fundamental principle of the PP technique involves initiating an optical process in the sample using a high-intensity pump pulse and probing the dynamics of the process with a weaker, time-delayed probe beam. **Figure 4c** illustrates a typical PP setup. An ultrashort laser pulse is split into pump and probe beams using a beam splitter. The pump beam typically has an intensity around ten times greater than the

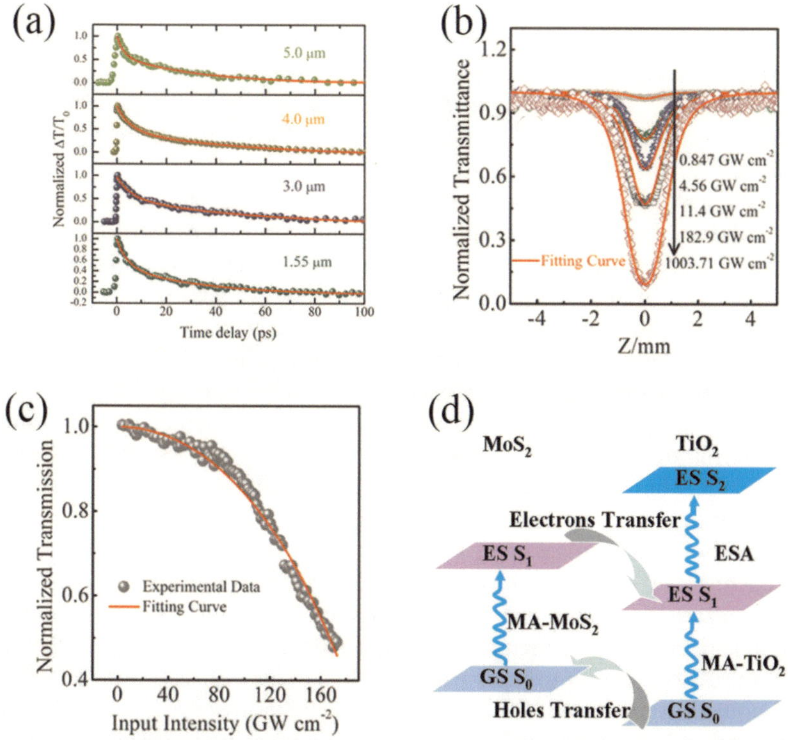

Figure 5.
(a) PP curves [23], (b) open-aperture Z-scan curves for optical limiting [14], (c) nonlinear transmission as a function of the input intensity [14], and (d) schematic diagram of the nonlinear absorption in MoS_2/TiO_2 nanocomposite [30].

probe beam. The time delay (Δt) between the two beams is controlled by an adjustable delay stage. Both beams are focused onto the sample, ensuring they overlap spatially. A typical PP curve is shown in the **Figure 5a**, and key information such as the recovery time can be obtained through curve fitting [23].

The pump beam excites the sample, altering its state properties, while the probe beam detects these perturbations. By varying the time delay between the pump and probe pulses, information about the decay dynamics of excited carriers and the evolution of the sample's excited state can be obtained.

PP spectroscopy can be implemented in various configurations, depending on the nature of the investigation. For instance, the pump and probe beams may have identical frequencies (degenerate PP) or different frequencies (nondegenerate PP). Transient absorption (TA) spectroscopy, a specific type of nondegenerate PP technique, uses a supercontinuum light source as the probe.

4. NLO characterization and application of TMDs

4.1 TMDs for optical limiting

The widespread use of lasers has driven significant advancements in laser protection for high-value optoelectronic devices, military equipment, and human safety.

However, traditional laser protection tools, such as protective glasses, have notable limitations [14, 27, 31]. These devices rely primarily on the linear absorption of light, which results in drawbacks like a narrow protection wavelength range, poor resistance to pulsed lasers, and a trade-off between protection and light transmission. Similarly, phase-transition materials like VO_2 and V_2O_5 offer slow response times, limiting their effectiveness.

In contrast, optical limiting based on third-order nonlinearity has emerged as a promising solution. This approach addresses the limitations of traditional methods by offering fast response times and high efficiency. Nonlinear optical limiters act as intelligent optical switches, allowing low-intensity light to pass while blocking harmful high-intensity light, thereby demonstrating self-adaptive behavior. Materials exhibiting RSA in the visible spectrum are particularly effective, as their excited-state absorption is significantly higher than that in the ground state.

Optical limiting relies on nonlinear optical phenomena such as multiphoton absorption, nonlinear refraction, and nonlinear scattering [14, 16, 28, 30]. To function effectively, optical limiting materials must exhibit the ability to absorb, refract, or scatter incident light through these nonlinear effects.

Figure 5b shows the typical open-aperture *Z*-scan curve for optical limiting [14]. As Z approaches 0 ($Z = 0$ being the focal point), the intensity of the pump light gradually increases, leading to enhanced absorption by the optical limiter and a reduction in transmittance. When moving to the focal point at $Z = 0$, the absorption is at its maximum (with minimum transmittance). Conversely, as Z moves farther from the focal point, the intensity of the pump light decreases, resulting in reduced absorption by the optical limiting material and increased transmittance. The curve obtained in **Figure 5a** can be fitted using Eq. (22). By combining Eqs. (23), (25), and (26), the β, $Re\chi^{(3)}$, $Im\chi^{(3)}$ and $\chi^{(3)}$ as well as n_2 can be determined. Another criterion for evaluating the application of optical limiter is the optical limiting threshold, which refers to the pump laser power density (or light intensity) corresponding to a transmittance that decreases to 50% of the initial transmittance [14], as shown in **Figure 5c**.

TMDs exhibit excellent optical limiting effects. Typically, when TMDs are pumped with a laser whose energy (E_{laser}) is lower than the bandgap energy (E_g), they exhibit strong optical limiting effects due to mechanisms such as reverse saturable absorption, multiphoton absorption, excited-state absorption, and free-carrier absorption. Unlike graphene's zero bandgap, TMDs typically have a bandgap ranging from 0.5 to 2.2 eV. This variation in bandgap caused by different elemental compositions broadens the application scope of TMDs in optical limiting. As a representative layered material, the bandgap of TMDs decreases, and their photoluminescence redshifts as the number of layers increases, leading to changes in the range of their optical limiting effects. Different sizes of TMDs, such as nanosheets and quantum dots, exhibit distinct nonlinear properties, meaning their optical limiting characteristics also vary. TMDs can be combined with other materials to achieve superior optical limiting performance through energy transfer mechanisms. For example, Rongfei Wei et al. reported a composite material of MoS_2 and TiO_2. Under strong light excitation, electrons in MoS_2 are transferred to TiO_2 through a photoinduced process [30], as shown in **Figure 5d**. This electron transfer provides additional electron storage sites for the composite material, thereby enhancing its optical limiting performance and significantly reducing the optical limiting threshold. This composite strategy leverages the synergistic effects of the two materials, effectively optimizing the optical limiting properties and expanding the application potential of TMDs in strong light protection fields. The obtained β for TMDs typically falls within the range of 10^{-11} to 10^{-10} m/W [8–10].

However, due to the resonance enhancement effect, optical limiting effects may still occur even when $E_{laser} > E_g$, rather than the expected SA effect (a detailed discussion of SA will be provided in the next section). For instance, Xiangling Tian and colleagues [14] prepared multilayer WSe_2 using CVD, which has a bandgap of 1.49 eV. When this multilayer WSe_2 was pumped with a 1.55 eV laser, the enhanced resonance phenomenon allowed electrons excited to the conduction band to continue absorbing energy and transition further upward with the assistance of phonons. This created "new" storage sites for electrons in the conduction band that would otherwise have been fully occupied, resulting in strong optical limiting effects. The optical limiting threshold observed for WSe_2 was remarkably low at 21 mJ/cm^2 at 800 nm (**Figure 5c**), significantly outperforming many previously reported materials. For comparison, the thresholds for other optical limiting materials are approximately 117 mJ/cm^2 for conjugated organic molecules, 26,000 mJ/cm^2 for metal nanoparticles, 74 mJ/cm^2 for carbon nanodots, 37 mJ/cm^2 for graphene oxide, and 44 mJ/cm^2 for MoS_2 nanosheets [27, 28, 32]. These results highlight the strong potential of WSe_2 for practical applications in femtosecond laser safety, such as protecting eyes and sensor focal-plane arrays.

Closed-aperture *Z*-scan measurements can be performed to obtain the closed-aperture curve, which, when divided by the open-aperture data, yields the normalized transmittance curve. By fitting this curve, the nonlinear refractive index n_2 can be determined. The value of n_2 typically varies with pump power and, when analyzed alongside the optical limiting threshold, provides a useful metric for evaluating the optical limiting effect.

4.2 SA and pulsed lasers

Another example of third-order nonlinearity is saturated absorption, which can be identified through open-aperture *Z*-scan measurements. The curve reveals that as the sample moves closer to the focal point (Z=0), the pump light intensity increases, leading to saturation of absorption and an enhanced transmittance. Conversely, as the sample moves away from the focal point, the pump light intensity decreases, and the absorption gradually returns to its initial state. On the open-aperture *Z*-scan curve, this phenomenon is represented by a peak, as shown in **Figure 6a**. The curve can be fitted using Eq. 22 to obtain the β. The β not only is related to the pump intensity but also depends on factors such as the number of layers, size, and type of material. For TMDs, the β varies widely depending on the preparation methods and research conducted by different groups. For instance, at 800 nm, the β ranges from 10^0 to 10^4 cm/GW. Detailed discussions can be found in some review articles, and specific numerical values will not be listed in this section.

As previously noted, an SA is essential as a pulse shaping device in both scenarios. In the past decade, TMDs-based SAs have found extensive applications in passively mode-locked and Q-switched lasers, leading to numerous significant advancements. Integrating TMDs into waveguide devices to form pulse shaping devices or optical modulators leverages the nonlinear properties of TMDs to shape the pulse. For instance, researchers can position TMDs between two optical connectors. In this approach, the materials are typically transferred to the surface of one connector using techniques such as optical deposition or polymer film transfer. Another proposed

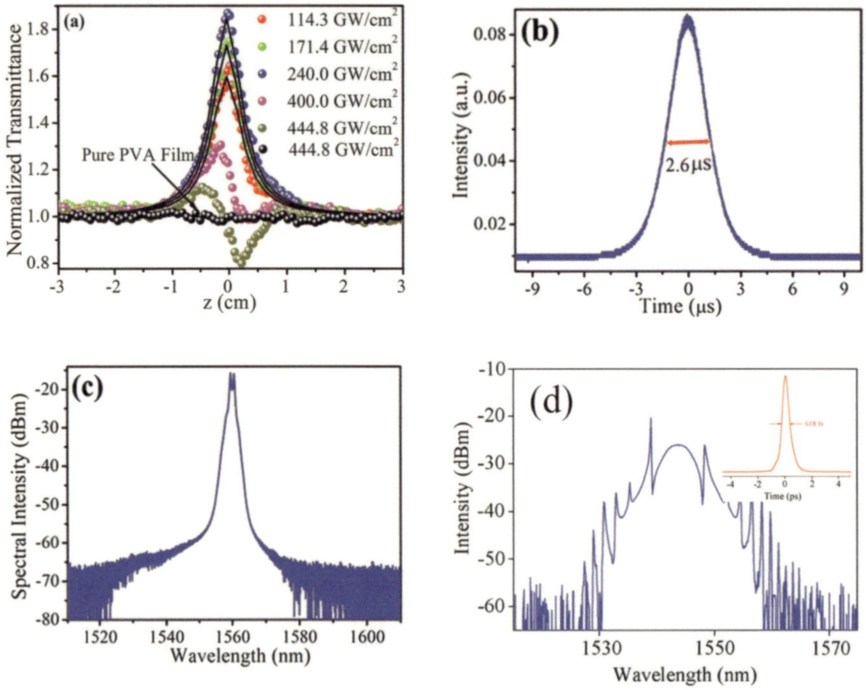

Figure 6.
(a) open-aperture Z-scan curves for saturable absorber [33], (b) single Q-switching pulse profile based on MoS_2 nanoflower [33], (c) output spectrum of the Q-switching pulse [33], and (d) Mode-locked optical spectrum and temporal width of one pulse [34].

method involves creating microfluidic channels within optical fibers. Alternatively, integrating layered materials into micro-structured optical fibers, such as photonic crystal fibers, D-shaped fibers, or tapered fibers, has also gained attention. Recently, fully integrated monolithic fiber lasers have emerged as a research focus. As micro- and nano-fabrication technologies advance, significant performance improvements are expected. For applications such as space transmission or solid-state lasers, layered materials are commonly deposited onto optical lenses via spin-coating or direct-dropping methods to create SA mirrors.

Since 2009, there has been rapid progress in the development of various pulsed lasers incorporating TMDs as saturable absorber. These include fiber lasers, solid-state lasers, disk lasers, and waveguide lasers, showcasing the versatility of this technology. This topic has been extensively reviewed in numerous articles, and this chapter will not repeat those discussions. To achieve such pulses, the primary mechanisms employed are Q-switching and mode-locking.

To achieve stable continuous-wave mode-locking without Q-switching instabilities, the condition $E_P^2 > E_{L,sat}E_{A,sat}\Delta m$ must be satisfied [33]. In this equation, E_P represents the intra-cavity pulse energy. Δm is the modulation depth of the saturable absorber. $E_{L,sat}$ and $E_{A,sat}$ denote the effective saturation energies of the gain medium and the saturable absorber, respectively. On the other hand, generating stable Q-switched pulses requires the saturable absorber to have both a high modulation depth and a large saturation energy. These characteristics help suppress mode-locking while maintaining precise control of the cavity dispersion.

For the spectrum of mode-locked pulses, the following two conditions must be satisfied [35]: $\sigma_A/A_A > \sigma_G/A_G$ and $\tau_A < \tau_G$, where σ_A and σ_G denote the effective cross sections of the absorber and the gain, respectively. A_A and A_G denote the geometrical cross section of the laser beam within the absorber and gain regions. τ_A and τ_G correspond to the lifetimes associated with saturation effects in the absorber and gain.

So far, saturable absorbers based on TMDs have achieved ultra-wide spectral pulse modulation in the range of ≈1000–2000 nm through Q-switching and mode-locking techniques [34]. The repetition frequency spans from 4.5 kHz to 216 MHz, with some pulses reaching several thousand MHz [36]. For instance, using MoS_2 nanosheets as the absorber, a mode-locked laser at 1558 nm can achieve a repetition frequency of 2500 MHz [37]. Similarly, employing $MoSe_2$ nanosheet films as the optical switch enables a mode-locked pulse laser at 1557.3 nm with a repetition frequency of 3270 MHz [37]. The signal-to-noise ratio (SNR) can be optimized to ≈25–100 dB, although the SNR of mode-locked pulses is typically higher than that of Q-switched pulses [37]. For mode-locked pulses, the pulse duration generally ranges from femtoseconds to picoseconds, with the shortest mode-locked pulse achieved using TMDs being 67 femtoseconds [38]. In contrast, Q-switched pulse widths are typically on the microsecond scale. The maximum pulse energy can reach 1179.4 nJ [39].

Generally, saturable absorbers with high modulation depth are conducive to achieving Q-switched pulse. For instance, Rongfei Wei et al. [33] reported MoS_2 nanoflower-based saturable absorber with a high modulation depth of 51.8% and a large saturable intensity of 275.5 GW/cm^2 was developed for use in passively Q-switched fiber laser systems, as shown in **Figure 6b, c**. The laser exhibited an efficiency of 4.71%, surpassing the performance of previously prepared layered MoS_2 and other recently reported MoS_2-based Q-switched lasers. This material is characterized by an extremely fast recovery time. A TiS_2-based saturable absorber exhibits nonlinear optical SA with a recovery time of approximately 768 fs, enabling modulation frequencies exceeding 1.3 THz [34]. In an all-optical system, it has been used to generate a stable femtosecond mode-locked pulse with a duration of around 402 fs and a pulse train with a period of approximately 175.5 nanoseconds, as shown in **Figure 6d**.

4.3 Optical modulators

Beyond their role as saturable absorbers for pulse laser generation, TMDs are also effective as optical modulators for optical modulation [40]. These devices physically or logically manipulate optical signals within transmission lines or integrated optical pathways, especially in high-speed, short-range communication systems. Optical modulators are pivotal for applications such as wavelength conversion, demultiplexing, THz wave generation, nonlinear holography, and optical cross-connections. Thanks to their nanoscale structure and phase-matching properties, 2D materials offer notable advancements in speed and energy efficiency.

THz radiation encompasses electromagnetic waves with frequencies between 0.1 and 10 THz, corresponding to wavelengths from 30 μm to 3 mm. This radiation is widely utilized in medical diagnosis, optical imaging, and communication technologies. In 2018, Xu et al. [41] explored the THz radiation emitted by layered WSe_2 under ultrafast laser excitation, attributing it to the *d-d* photo-transition mechanism. Their findings also revealed that WSe_2 has a laser damage threshold of 3.11 mJ cm^{-2}, underscoring its suitability for THz generation.

Nonlinear holography, which records and reproduces 3D images at new frequencies by combining holography with NLO, has applications in optical communication, data storage, and security. Gao et al. [42] reported a high-efficiency nonlinear hologram using a single nano-patterned WS_2 monolayer under second-harmonic frequency excitation. In the SHG range of 400–520 nm, the device exhibited significantly higher conversion efficiency than earlier plasmonic metasurfaces. This nonlinear polarization showed directional selectivity in both horizontal and vertical orientations, paving the way for the integration of nonlinear nano-holography into optical imaging, data storage, and secure electronic devices.

All-optical phase shifters and switches are essential for various optical applications, including signal processing, sensing, and communication. Wu *et al.* demonstrated a fiber-based all-optical phase shifter utilizing few-layer WS_2 deposited on a tapered fiber [43]. By absorbing 980 nm pump light, WS_2 generates heat, altering the refractive index of both the material and the fiber *via* the thermo-optic effect, achieving a maximum phase shift of 6.1π near 1550 nm. This cost-effective, compact device showcases TMDs' potential in all-optical signal processing. However, challenges such as inefficient light–matter interaction due to the subnanometer thickness of 2D materials and optical damage caused by high excitation power must be addressed to fully harness their nonlinear properties. To overcome these limitations, heterostructures of 2D materials [44, 45], such as MoS_2–WS_2 or MoS_2–WSe_2, have been proposed. These designs exhibit tunable optical properties, including reflectance and carrier dynamics, but further exploration is needed to realize their full potential in nonlinear optics.

SHG has proven valuable for identifying the layer number and orientation of TMDs. Additionally, nonlinear frequency conversion processes, such as four-wave mixing and harmonic generation reversal, are key methods for producing quantum light sources. The high optical nonlinearity of TMDs-based materials holds promise for creating high-purity quantum emitters and switches in integrated quantum circuits with atomic-scale thickness.

4.4 Nonlinear optical microscopies

By integrating the unique characteristics of spectroscopy and microscopy, dynamic biological processes, such as those involving nucleic acids, proteins, and lipids, can be observed directly and without causing damage. Innovative tools in microbiology, like nonlinear optical microscopes (NOMs), have significantly advanced biological sciences. Over the past decade, nonlinear optical techniques have gained prominence in detecting biomolecules at micro- and nanoscale levels, as well as in applications across medicine and pharmacy. Optical methods offer noninvasive, real-time, precise, and cost-effective molecular insights into human tissues during disease detection, diagnosis, and cellular development.

4.4.1 Biological applications of SHG

Interfacial biomolecular activities are fundamental to life processes. Understanding the dynamics of biomolecules at interfaces is critical for constructing accurate disease models, enhancing drug screening, and deciphering spatiotemporal energy and matter transport in living systems. Cui et al. [46] demonstrated a label-free sensing method by observing biomolecular adsorption and desorption on monolayer MoS_2 surfaces using time-resolved SHG. This approach utilized chitosan nanoclusters and bovine serum albumin (BSA) molecules, which formed electrostatic attractions with the negatively charged MoS_2 surface. By tracking changes in SHG intensity over time, biomolecular interactions in liquid environments were detected in real time. Furthermore, SHG emission from MoS_2 was manipulated using single-cell bacteria, enabling anisotropic polarization dependence and the development of label-free single-bacterium sensors. A transient SHG signal change caused by UV-induced bacterial cell-wall rupture enabled identification of individual bacterial cells within a mixed population [47].

4.4.2 Biological applications of TPEE

TPEE offers advantages over single-photon microscopy, such as reduced background fluorescence in living tissues, decreased photodamage and bleaching of biological samples, enhanced 3D spatial resolution, and deeper tissue penetration. A noteworthy example involves the creation of a water-soluble, antibody-conjugated MoS_2 quantum dot (QD)-based two-photon luminescence (TPL) probe for targeted cancer cell imaging in the biological II window [48]. MoS_2 QDs exhibit an exceptionally high two-photon absorption cross section (σ = 58,960 GM) and brightness (4.7×10^3 GM) due to quantum confinement and edge effects. Experimental results demonstrated the selective imaging of live prostate cancer cells using anti-PSMA antibody-attached MoS_2 QDs and 1064 nm light. These QDs exhibited excellent photostability, biocompatibility, and the ability to distinguish targeted from nontargeted cells, making them an efficient agent for high-precision live-cell imaging.

4.4.3 Biological applications of SERS

Surface-enhanced Raman spectroscopy (SERS) enhances the Raman signals of molecules, providing a highly sensitive method for detecting even single molecules [49]. A novel van der Waals heterostructure substrate combining TMD nanodomes (MoS_2 and WS_2) with graphene demonstrated extraordinary SERS sensitivity. Using fluorescent Rhodamine 6G (R6G) molecules as probes, this substrate achieved detection limits between 10^{-11} and 10^{-12} M with 532 nm Raman excitation [50]. This performance surpasses single-layer TMD or graphene substrates by over three orders of magnitude and rivals the best results obtained with plasmonic metal nanostructures.

5. Conclusion and outlook

This chapter presents a thorough examination of the fundamental theories, characterization methodologies, and prospective application of TMDs within the field of NLO. TMDs exhibit exceptional broadband absorption and notable NLO responses, indicating significant potential for applications in optical limiting, saturable absorption, pulsed laser generation, and biological sensors.

The analysis of physical mechanisms underpinning second- and third-order NLO processes underscores the importance of experimental techniques, such as Z-scan, pump-probe spectroscopy, and multiphoton microscopy imaging for characterizing the optical properties of NLO materials. Furthermore, the utilization of TMDs in cutting-edge technologies such as optical modulators, terahertz radiation generation, and nonlinear holography underscores their versatility in optoelectronic devices.

In the biomedical sector, the nonlinear optical properties of TMDs have been effectively leveraged in advanced techniques such as two-photon luminescence imaging and SERS. These applications offer innovative solutions for noninvasive, high-precision biological detection and diagnosis. Nevertheless, challenges persist, including low light-matter interaction efficiency and the potential for laser-induced optical damage. The development of heterostructures and novel composite materials may further enhance the performance of TMDs and unlock their full potential in optical and optoelectronic applications.

In conclusion, TMDs, characterized by their unique physical and optical properties, not only contribute to the advancement of fundamental research in nonlinear optics but also present promising applications, thereby establishing a robust foundation for the development of next-generation photonic and optoelectronic technologies.

Acknowledgements

This work was supported by the National Natural Science Foundation of China (Grant No. 12204435), the Natural Science Foundation of Zhejiang Province (Grant No. LQ22A040011), and the Municipal Government of Quzhou (Grant Nos. 2023D012).

Conflict of interest

The authors declare no conflict of interest.

Author details

Xiangling Tian[1*], Lun Wang[1,2] and Rongfei Wei[3]

1 Yangtze Delta Region Institute (Quzhou), University of Electronic Science and Technology of China, Quzhou, Zhejiang, China

2 The College of Computer Science and Technology, Xinjiang University, Urumqi, China

3 Department of Physics, Zhejiang Normal University, Jinhua, Zhejiang, China

*Address all correspondence to: xianglingt@sina.com and rfwei@zjnu.cn

References

[1] Gupta MC, Ballato J. The Handbook of Photonics. Boca Raton: CRC press; 2018. DOI: 10.1201/9781315222103

[2] Sauter E. Nonlinear Optics. New York: John Wiley & Sons; 1996

[3] Boyd RW, Gaeta AL, Giese E. Nonlinear optics. In: Springer Handbook of Atomic, Molecular, and Optical Physics. New York: Springer; 2008. pp. 1097-1110

[4] Franken PA, Hill AE, Peters CW, Weinreich G. Generation of optical harmonics. Physical Review Letters. 1961;7:118-119. DOI: 10.1103/PhysRevLett.7.118

[5] Kauranen M, Zayats AV. Nonlinear plasmonics. Nature Photonics. 2012;**6**: 737-748. DOI: 10.1038/nphoton.2012.244

[6] Choi W, Choudhary N, Han GH, Park J, Akinwande D, Lee YH. Recent development of two-dimensional transition metal dichalcogenides and their applications. Materials Today. 2017; **20**:116-130. DOI: 10.1016/j.mattod.2016.10.002

[7] Bloembergen N. Nonlinear spectroscopy. In: Laser Spectroscopy: Proceedings of the Second International Conference, Megève, June 23–27, 1975. Berlin, New York: Springer; 1975. pp. 31-38

[8] Reshef O, De Leon I, Alam MZ, Boyd RW. Nonlinear optical effects in epsilon-near-zero media. Nature Reviews Materials. 2019;**4**:535-551. DOI: 10.1038/s41578-019-0120-5

[9] Wen X, Gong Z, Li D. Nonlinear optics of two-dimensional transition metal dichalcogenides. InfoMat. 2019;**1**: 317-337. DOI: 10.1002/inf2.12024

[10] Ahmed S, Jiang X, Wang C, Kalsoom UE, Wang B, Khan J, et al. An insightful picture of nonlinear photonics in 2D materials and their applications: Recent advances and future prospects. Advanced. Optical Materials. 2021;**9**: 2001671. DOI: 10.1002/adom.202001671

[11] Tian X, Luo H, Wei R, Liu M, Yang Z, Luo Z, et al. Ultrafast and broadband optical nonlinearity in aluminum doped zinc oxide colloidal nanocrystals. Nanoscale. 2019;**11**: 13988-13995. DOI: 10.1039/C9NR04337C

[12] Denk W, Strickler JH, Webb WW. Two-photon laser scanning fluorescence microscopy. Science. 1990;**248**:73-76. DOI: 10.1126/science.2321027

[13] Sheik-Bahae M, Said AA, Van Stryland EW. High-sensitivity, single-beam n_2 measurements. Optics Letters. 1989;**14**:955-957. DOI: 10.1364/OL.14.000955

[14] Tian X, Wei R, Guo Q, Zhao Y-J, Qiu J. Reverse Saturable absorption induced by phonon-assisted anti-stokes processes. Advanced Materials. 2018;**30**: 1801638. DOI: 10.1002/adma.201801638

[15] Zaręba JK, Nyk M, Samoć M. Nonlinear optical properties of emerging Nano- and Microcrystalline materials. Advanced Optical Materials. 2021;**9**: 2100216. DOI: 10.1002/adom.202100216

[16] Tian X, Wei R, Yang D, Qiu J. Paradoxical combination of saturable absorption and reverse-saturable absorption in plasmon semiconductor nanocrystals. Nanoscale Advances. 2020; **2**:1676-1684. DOI: 10.1039/C9NA00694J

[17] Wei R, Qiao T, Tian X, Zhang H, He X, Hu Z, et al. Enhanced nonlinear

optical response of Se-doped MoS_2 nanosheets for passively Q-switched fiber laser application. Nanotechnology. 2017;**28**:215206. DOI: 10.1088/1361-6528/aa6bc8

[18] Wei R, Zhang H, Hu Z, Qiao T, He X, Guo Q, et al. Ultra-broadband nonlinear saturable absorption of high-yield MoS_2 nanosheets. Nanotechnology. 2016;**27**:305203. DOI: 10.1088/0957-4484/27/30/305203

[19] Tian X, Wei R, Liu S, Zhang Y, Qiu J. Photoluminescence nonuniformity from self-seeding nuclei in CVD-grown monolayer $MoSe_2$. Nanoscale. 2018;**10**: 752-757. DOI: 10.1039/C7NR08662H

[20] Khan AR, Zhang L, Ishfaq K, Ikram A, Yildrim T, Liu B, et al. Optical harmonic generation in 2D materials. Advanced Functional Materials. 2022;**32**: 2105259. DOI: 10.1002/adfm.202105259

[21] Ma L, Wei R, Yu Q, Dai P, Tian X, Hu F, et al. Improved broadband luminescence in $Gd_2GaSb_{1-x}Ta_xO_7$:Cr^{3+}, Yb^{3+} pyrochlore phosphors: Near-infrared spectroscopic applications and dual-mode optical thermometry. Materials Today Chemistry. 2024;**38**: 102090. DOI: 10.1016/j.mtchem.2024.102090

[22] Wei R, Tian X, Luo H, Liu M, Yang Z, Luo Z, et al. Heavily doped semiconductor colloidal nanocrystals as ultra-broadband switches for near-infrared and mid-infrared pulse lasers. ACS Applied Materials & Interfaces. 2019;**11**:40416-40423. DOI: 10.1021/acsami.9b10949

[23] Tian X, Luo H, Wei R, Zhu C, Guo Q, Yang D, et al. An ultrabroadband mid-infrared pulsed optical switch employing solution-processed bismuth oxyselenide. Advanced Materials. 2018;**30**:1801021. DOI: 10.1002/adma.201801021

[24] Wei R, Tian X, Yang L, Yang D, Ma Z, Guo H, et al. Ultrafast and large optical nonlinearity of a $TiSe_2$ saturable absorber in the 2 μm wavelength region. Nanoscale. 2019;**11**:22277-22285. DOI: 10.1039/C9NR06374A

[25] Sheik-Bahae M, Hutchings D, Hagan DJ, Van Stryland E. IEEE Journal of Quantum Electronics. 1991;**27**:1296

[26] de Araújo CB, Gomes ASL, Boudebs G. Techniques for nonlinear optical characterization of materials: A review. Reports on Progress in Physics. 2016;**79**:036401. DOI: 10.1088/0034-4885/79/3/036401

[27] Wei R, Tian X, Zhang H, Hu Z, He X, Chen Z, et al. Facile synthesis of two-dimensional WS_2 with reverse saturable absorption and nonlinear refraction properties in the PMMA matrix. Journal of Alloys and Compounds. 2016;**684**:224-229. DOI: 10.1016/j.jallcom.2016.05.169

[28] Wei R, Zhang H, He X, Hu Z, Tian X, Xiao Q, et al. Versatile preparation of ultrathin MoS_2 nanosheets with reverse saturable absorption response. Optical Materials Express. 2015;**5**: 1807-1814. DOI: 10.1364/OME.5.001807

[29] Ramteke SP, Baig MI, Shkir M, Kalainathan S, Shirsat MD, Muley GG, et al. Novel report on SHG efficiency, Z-scan, laser damage threshold, photoluminescence, dielectric and surface microscopic studies of hybrid inorganic ammonium zinc sulphate hydrate single crystal. Optics & Laser Technology. 2018;**104**:83-89. DOI: 10.1016/j.optlastec.2018.02.018

[30] Wei R, Tian X, Hu Z, Zhang H, Qiao T, He X, et al. Vertically standing layered MoS_2 nanosheets on TiO_2 nanofibers for enhanced nonlinear

optical property. Optics Express. 2016; **24**:25337-25344. DOI: 10.1364/OE.24.025337

[31] Shang W, Wang B, Wang Q. 2D materials for nonlinear optical limiting. In: Two-Dimensional Materials for Nonlinear Optics. Optics. Online books: Wiley; 2024. pp. 185-219. DOI: 10.1002/9783527838288.ch8

[32] Polavarapu L, Venkatram N, Ji W, Xu Q-H. Optical-limiting properties of oleylamine-capped gold nanoparticles for both femtosecond and nanosecond laser pulses. ACS Applied Materials & Interfaces. 2009;**1**:2298-2303. DOI: 10.1021/am900442u

[33] Wei R, Zhang H, Tian X, Qiao T, Hu Z, Chen Z, et al. MoS_2 nanoflowers as high performance saturable absorbers for an all-fiber passively Q-switched erbium-doped fiber laser. Nanoscale. 2016;**8**:7704-7710. DOI: 10.1039/C5NR08457A

[34] Tian X, Wei R, Liu M, Zhu C, Luo Z, Wang F, et al. Ultrafast saturable absorption in TiS_2 induced by non-equilibrium electrons and the generation of a femtosecond mode-locked laser. Nanoscale. 2018;**10**:9608-9615. DOI: 10.1039/C8NR01573B

[35] Smith PW, Silberberg Y, Miller DAB. Mode locking of semiconductor diode lasers using saturable excitonic nonlinearities. Journal of the Optical Society of America B: Optical Physics. 1985;**2**:1228-1236. DOI: 10.1364/JOSAB.2.001228

[36] Guo B, Xiao Q-L, Wang S-H, Zhang H. 2D layered materials: Synthesis, nonlinear optical properties, and device applications. Laser & Photonics Reviews. 2019;**13**:1800327. DOI: 10.1002/lpor.201800327

[37] Yu S, Wu X, Wang Y, Guo X, Tong L. 2D materials for optical modulation: Challenges and opportunities. Advanced Materials. 2017; **29**:1606128. DOI: 10.1002/adma.201606128

[38] Liu W, Pang L, Han H, Liu M, Lei M, Fang S, et al. Tungsten disulfide saturable absorbers for 67 fs mode-locked erbium-doped fiber lasers. Optics Express. 2017;**25**:2950-2959. DOI: 10.1364/OE.25.002950

[39] Chen B, Zhang X, Wu K, Wang H, Wang J, Chen J. Q-switched fiber laser based on transition metal dichalcogenides MoS_2, $MoSe_2$, WS_2, and WSe_2. Optics Express. 2015;**23**: 26723-26737. DOI: 10.1364/OE.23.026723

[40] Sun Z, Martinez A, Wang F. Optical modulators with 2D layered materials. Nature Photonics. 2016;**10**:227-238. DOI: 10.1038/nphoton.2016.15

[41] Si K, Huang Y, Zhao Q, Zhu L, Zhang L, Yao Z, et al. Terahertz surface emission from layered semiconductor WSe_2. Applied Surface Science. 2018; **448**:416-423. DOI: 10.1016/j.apsusc.2018.04.117

[42] Dasgupta A, Gao J, Yang X. Atomically thin nonlinear transition metal dichalcogenide holograms. Nano Letters. 2019;**19**:6511-6516. DOI: 10.1021/acs.nanolett.9b02740

[43] Wu K, Guo C, Wang H, Zhang X, Wang J, Chen J. All-optical phase shifter and switch near 1550nm using tungsten disulfide (WS_2) deposited tapered fiber. Optics Express. 2017;**25**:17639-17649. DOI: 10.1364/OE.25.017639

[44] Lee C-H, Lee G-H, van der Zande AM, Chen W, Li Y, Han M, et al. Atomically thin p–n junctions with van der Waals heterointerfaces. Nature

Nanotechnology. 2014;**9**:676-681. DOI: 10.1038/nnano.2014.150

[45] Rigosi AF, Hill HM, Li Y, Chernikov A, Heinz TF. Probing interlayer interactions in transition metal dichalcogenide heterostructures by optical spectroscopy: MoS_2/WS_2 and $MoSe_2/WSe_2$. Nano Letters. 2015;**15**: 5033-5038. DOI: 10.1021/acs.nanolett.5b01055

[46] Xia C, Sun J, Wang Q, Chen J, Wang T, Xu W, et al. Label-free sensing of biomolecular adsorption and desorption dynamics by interfacial second harmonic generation. Biosensors. 2022;**12**:1048. DOI: 10.3390/bios12111048

[47] Kim YC, Jun SW, Ahn YH. Single bacteria identification with second-harmonic generation in MoS_2. Biosensors and Bioelectronics. 2023;**241**:115675. DOI: 10.1016/j.bios.2023.115675

[48] Sweet C, Pramanik A, Jones S, Ray PC. Two-photon fluorescent molybdenum disulfide dots for targeted prostate cancer imaging in the biological II window. ACS Omega. 2017;**2**: 1826-1835. DOI: 10.1021/acsomega.7b00229

[49] Ray PC. Size and shape dependent second order nonlinear optical properties of nanomaterials and their application in biological and chemical sensing. Chemical Reviews. 2010;**110**:5332-5365. DOI: 10.1021/cr900335q

[50] Kaur J, Singh M, Aversana CD, Benedetti R, Giardina P, Rossi M, et al. Biological interactions of biocompatible and water-dispersed MoS_2 nanosheets with bacteria and human cells. Scientific Reports. 2018;**8**:16386. DOI: 10.1038/s41598-018-34679-y

Chapter 7

MoS_2 for Cancer Diagnosis and Therapy

Wanting Yang and Wensheng Xie

Abstract

Molybdenum disulfide (MoS_2), a typical two-dimensional (2D) transition metal dichalcogenide (TMDs), has gained significant interest in cancer diagnosis and therapy due to its unique properties, such as excellent photothermal conversion efficiency, high surface area, ease of functionalization, and biocompatibility. Meanwhile, the easily controllable preparation and precise synthesis of MoS_2 nanomaterials can be achieved according to the needs of cancer diagnosis and treatment. In this chapter, we introduced the latest developments of MoS_2 nanomaterials in cancer diagnosis and therapy, mainly focusing on photothermal therapy (PTT), microwave thermal therapy (MWT), drug delivery system (DDS), and cancer imaging. Furthermore, we also discuss the current challenges and potential prospects of MoS_2 nanomaterials for cancer diagnosis and therapy in the near future, providing guidelines for the novel design of MoS_2 nanomaterials to improve the therapeutic efficacy of cancer.

Keywords: MoS_2, photothermal therapy, microwave thermal therapy, drug delivery system, cancer imaging

1. Introduction

Transition metal dichalcogenides (TMDs) are a class of two-dimensional (2D) materials characterized by an X-M-X layered structure, where M represents a transition metal (Mo, W, V, Ti, etc.) and X denotes a chalcogenide (S, Se, or Te) [1]. Within a single layer, the atoms X-M-X are covalently bonded, while the individual layers are held together by weak Van der Waals forces. TMDs exhibit exceptional photothermal conversion efficiency and possess a substantial surface area, making them highly suitable for the targeted delivery of biomolecules and pharmaceuticals to specific cells, thereby enhancing the efficacy of cancer imaging and treatment [2]. These TMDs can also be utilized in a wide range of fields including sensors, electrodes, supercapacitors, batteries, optoelectronic components, catalysts, and spintronic systems.

Molybdenum disulfide (MoS_2), a prototypical TMD, is composed of hexagonal layers. Each layer, known as a "single layer" or "monolayer," consists of a three-atom-thick S-Mo-S stack, with a layer of molybdenum (Mo) atoms intercalated between two layers of sulfur (S) atoms through covalent bonding [3]. MoS_2 presents much polytypism with four different polytypes format, namely 1 T (trigonal), 2 T, 2H (hexagonal), and 3R (rhombohedral) (**Figure 1**). Meanwhile, researchers have synthesized various morphology and allotropies, including zero-dimensional

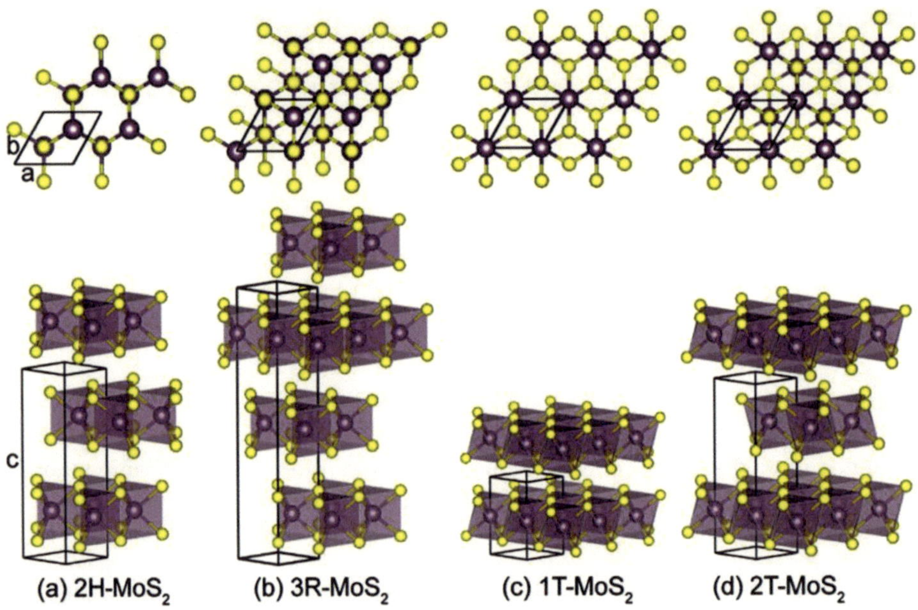

Figure 1.
Structural polytypes of layered MoS_2. Permission adapted from [4].

(nanoplatelets), one-dimensional (nanorods and nanowires), two-dimensional (nanosheets, nanoribbons, and nanostripes), and three-dimensional (nanoflowers and dandelion).

MoS_2 are considered biocompatible and safe due to the essential roles of both Mo and S atoms. Mo is a trace dietary element necessary for the survival of humans, participating in various metabolic processes through Mo-bearing enzymes. Meanwhile, S is a key component of organic compounds in living cells and is relatively abundant in the human body [4]. The biocompatibility of MoS_2 is further supported by studies showing minimal cytotoxicity even at high concentrations and its ability to be internalized by immune cells and cancer cells without causing significant harm. The favorable biocompatibility of MoS_2 has made them promising candidates for biomedical applications.

Besides the biocompatibility and minimal cytotoxicity, MoS_2 also possesses the following characteristics: photoluminescence (PL) arising from an indirect bandgap, unique fluorescence features; notable photothermal attributes, like robust absorption in the near-infrared (NIR) region; exceptional photothermal conversion efficiency; catalytic activity resembling enzymes; and significant X-ray attenuation capacity due to the high atomic number of Mo. In recent years, cancer treatment has evolved beyond traditional approaches like surgery, chemotherapy, and radiotherapy, which are often associated with significant side effects and the development of drug resistance. To address these limitations, emerging therapies such as photothermal therapy (PTT), photodynamic therapy (PDT), and microwave (MW) thermal therapy have gained increasing attention. The properties of MoS_2 provide MoS_2 remarkable performances in cancer PTT, MW thermal therapy, drug delivery system (DDS), and imaging. In this chapter, the advances of MoS_2 nanomaterials for cancer diagnosis and therapy will be systematically summarized and discussed, aiming to provide guidelines for the novel design of MoS_2 nanomaterials to improve the therapeutic efficacy of cancer.

2. MoS_2 for cancer photothermal therapy

Bulk MoS_2 has an indirect bandgap (1.2 eV) while the exfoliated MoS_2 exhibits direct bandgap (1.8 eV) and thus shows absorption and photoluminescence properties. The strong photoluminescence of monolayer MoS_2 makes it an attractive material for light-emitting devices and sensors. Its direct bandgap facilitates efficient photon absorption, which is crucial for converting light energy into heat, a property that is highly desirable in PTT for cancer treatment. Additionally, MoS_2 exhibits excellent optical absorption and photothermal conversion efficiency in the NIR window, facilitating NIR-based phototherapy and controlled drug release.

Changhui Fu et al. proposed the utilization of interlayer-expanded MoS_2 (E-MoS_2) nanosheets as potent agents for converting light energy into heat in the context of PTT [5]. The thin profile and increased interlayer gaps (around 0.94 nm) of these nanosheets conferred a notably high photothermal conversion efficiency, attributed to the enhanced absorption of near-infrared radiation. The obtained E-MoS_2 nanosheets also exhibited good biocompatibility and significant photothermal therapeutic effects. More importantly, E-MoS_2 nanosheets have been proven to be very effective for PTT in cancer cells and tumor-bearing mice. This study introduced a novel strategy for boosting the power conversion efficiency (PCE) of 2D layered nanomaterials by increasing the interlayer spacing. Liang Chen et al. combined bovine serum albumin-gadolinium (BSA-Gd) complex with MoS_2 nanoflakes to form a MoS_2-Gd-BSA complex, which can serve as a theranostic agent for both magnetic resonance (MR) and photoacoustic (PA) dual-modality imaging, as well as a therapeutic method for PTT [6]. This approach used a mild biomineralization method to synthesize BSA-Gd complexes. The MoS_2 nanoflakes were then surface-coated with poly(allylamine hydrochloride)(PAH) and poly(acrylic acid) (PAA) layers. Subsequently, the conjugation of BSA-Gd complexes with MoS_2 nanoflakes (denoted as MoS_2-Gd-BSA) was achieved through the amine reaction between the amino groups of the BSA-Gd complexes and the carboxyl groups present on the MoS_2 nanoflakes. Within this framework, the embedded MoS_2 nanoflakes served as a potent agent for PA and PTT, whereas the dual-layer PAH/PAA and the outer BSA-Gd functioned as the reactive bridge and MR imaging contrast medium, respectively. The newly synthesized MoS_2-Gd-BSA demonstrated superior photostability and strong photothermal response. Some studies modified MoS_2 nanosheets by PEGylation to enhance the therapeutic effect. Teng Liu et al. developed a type of MoS_2 nanosheets that have been chemically exfoliated and functionalized with polyethylene glycol (PEG) terminated with lipoic acid (LA-PEG), known as PEGylated MoS_2 (MoS_2-PEG) [7]. This modification endowed the MoS_2-PEG with high stability in biological environments and minimal toxicity. Capitalizing on its large surface area, the MoS_2-PEG could adsorb a photodynamic therapy agent named chlorin e6 (Ce6) through physical adsorption. The research revealed that MoS_2-PEG can efficiently transport Ce6 molecules into cells, enhancing the efficacy of photodynamic therapy in cancer cell elimination. This work suggested that transition metal dichalcogenides (TMDCs) with suitable surface coatings, such as the MoS_2-PEG proposed in this study, could become another class of 2D nanomaterials with great potential for applications in the biomedical field. Liang Chen et al. proposed a one-pot hydrothermal synthesis of MoS_2 nanoflakes with ideal degradability. Incorporating PAA into the hydrothermal process not only promotes the alteration of PEG but also endows them with the capacity to degrade [8]. MoS_2 – PPEG exhibited varying rates of degradation under different conditions, with swift decomposition in neutral pH solutions and significantly slower decay in the acidic settings typical of tumor microenvironments. And the results showed that the primary product of MoS_2 – PPEG degradation was water-soluble molybdenum ion. Furthermore, thanks to its superior photothermal properties, MoS_2 – PPEG was capable of effectively eliminating cancer cells *in vitro* and inhibiting tumor growth *in vivo*. The favorable *in vitro* biocompatibility of MoS_2 – PPEG was validated through assessments of cytotoxicity and hemolysis. Thus, the biodegradable MoS_2 nanoflakes synthesized *via* one-pot hydrothermal process could offer valuable insight for the future biomedical utilization of inorganic photothermal agents. Wensheng Xie et al. successfully designed a core/shell $Fe_3O_4@MoS_2$ nanocube (NC) structure (**Figure 2**), known as IOMS NCs, for MRI-directed chemo-photothermal treatment of human breast cancer xenografts in nude mice [9]. This was achieved by depositing a MoS_2 (MS) film onto iron oxide (IO) nanocubes using a simple hydrothermal synthesis approach. To enhance circulation time, the NCs underwent PEGylation and were subsequently additionally modified with 2-deoxy-D-glucose (2-DG) to enhance the concentration of the prepared NCs within the tumor region due to its appeal to cancer cells. These IOMS-PEG-2DG NCs served as carriers for DOX, which could be readily encapsulated within the NCs. The resulting IOMS-PEG(DOX)-2DG NCs displayed a good T2 relaxivity coefficient and superior photothermal capabilities. Following the intravenous administration of IOMS-PEG(DOX)-2DG NCs, the tumor location was distinctly identified 24 hours later through the intensified T2-weighted MRI signal. After being irradiated with an NIR

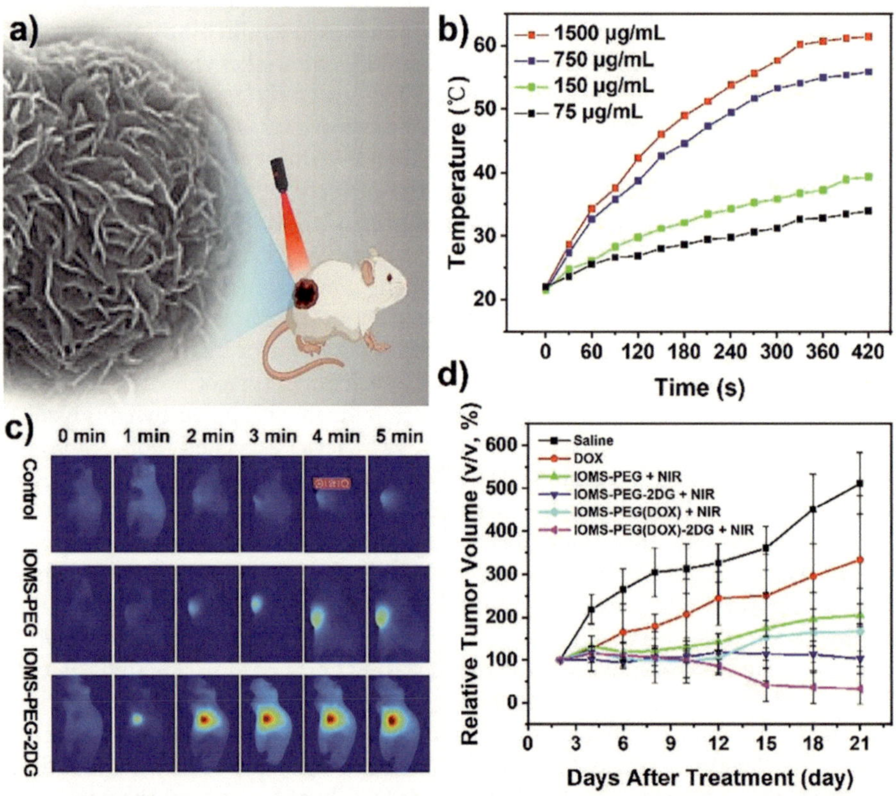

Figure 2.
Cancer phototherapy therapy using MoS_2-based nanomaterials. Permission adapted from [10].

808-nm laser for 5 minutes at a minimal power density, there was a significant rise in temperature at the tumor site, leading to effective suppression of tumor growth. In summary, the research underscored the effectiveness of IOMS-PEG(DOX)-2DG NCs as a superior theranostic agent, demonstrating promise for targeted and precise MRI-guided chemo-photothermal cancer treatment.

In summary, the 2D nanomaterial MoS_2 possesses excellent photothermal conversion performance, chemical inertness, and a vast potential for surface modification, which endows it with significant research potential in the field of photothermal therapy for tumors. With the continued advancement of nanobiomedicine, we believe that MoS_2 targeting modification materials with even more excellent photothermal conversion performance will be applied in the process of photothermal synergistic therapy for cancer, which will provide new ideas for exploring cancer treatment methods.

3. MoS_2 for cancer microwave thermal therapy

The interlayer distance in MoS_2 typically measures either 6.3 Å or 2.7 Å, providing a suitable space for the intercalation of ions such as H^+ or Li^+ [10]. This intercalation capability implies that there is a compartment between the layers where ions or molecules can be retained. Concurrently, microwave (MW) induces heating primarily through dipolar polarization and ionic conduction. Exposure to MW radiation prompts the alignment of dipoles or ions within the alternating electric field, leading to the generation of heat through molecular friction and energy loss due to dielectric properties. However, the intense heat produced by MW can potentially cause severe damage to healthy tissues and result in unbearable pain. Fortunately, the detrimental impacts of MW have been significantly mitigated by nanomaterials that respond to MW-induced heating. The sandwich structure of nanometer scale-spaces in MoS_2 can markedly enhance the heating effect, which promotes its application in MW hyperthermia.

Shengping Wang et al. prepared layered MoS_2 nanoflowers through a simple bottom-up hydrothermal process [10]. When these MoS_2 nanoflowers were encapsulated with BSA (resulting in BSA-MoS_2), they demonstrated minimal biological toxicity and superior absorption of MW energy *in vitro*. Additionally, the enhanced heating performance due to the layered architecture was validated through computational simulation. Ultimately, the stratified BSA-MoS_2 nanoflowers were engineered to serve as MW hyperthermia agents for *in vivo* cancer treatment, achieving complete tumor eradication upon exposure to MW radiation at a power of 1.8 W and a frequency of 450 MHz. This study introduced layered MoS_2 nanoflowers as a novel class of nanoagents for MW-induced cancer thermotherapy, marking a pioneering application in this field. Changhui Fu et al. synthesized a potent MW embolization agent, MSMCs, by encapsulating MoS_2 nanosheets within sodium alginate microcapsules, which could be effectively used for the treatment of large tumors (**Figure 3**) [11]. The biocompatibility assessment revealed that the MSMCs exhibited favorable non-toxicity *in vitro*. MSMCs were well distributed around the tumors, aiding in the embolization of artery vessels feeding the tumors and improving the therapeutic effect of MW ablation. Therefore, MSMC was a superior embolic and MW-susceptible agent, suitable for enhanced MW ablation treatment. The combination of MW ablation and TAE in the rabbit liver VX2 tumor model resulted in a significantly larger ablation zone, highlighting the synergistic effect of MSMC and making it a very promising

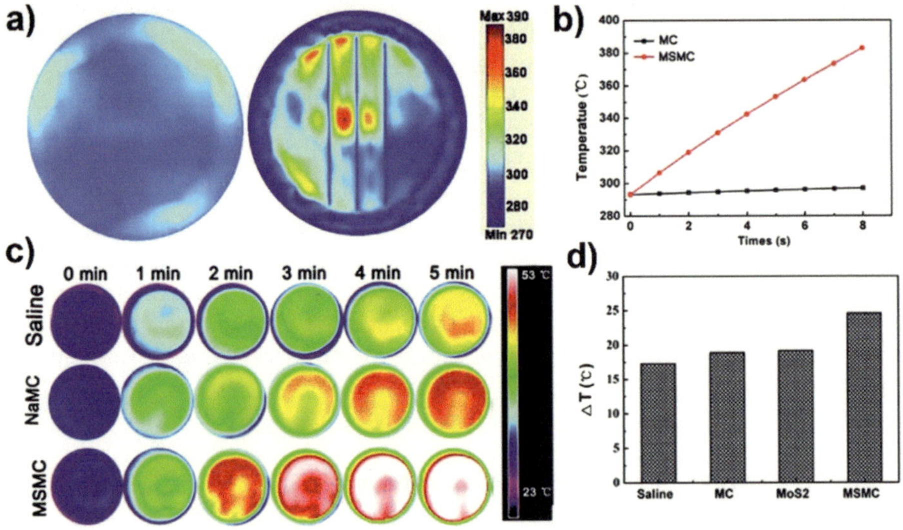

Figure 3.
Cancer microwave therapy using MoS_2-based nanosheets. Permission adapted from [12].

multifunctional therapeutic agent. A strategy utilizing MW technology to combat *Staphylococcus aureus*-induced osteomyelitis has been reported [12]. This approach involved the application of an MW-responsive composite made from MoS_2 and ferrous sulfide (FeS), which formed a heterojunction with the anti-inflammatory herb rhein (Rhe) to effectively kill bacteria. The polarization due to dipole alignment, which arises from the interfacial polarization in the MoS_2/FeS composite, is the primary driver of MW catalysis. Due to the anti-inflammatory properties of Rhe and the intensified MW thermal and kinetic effects of MoS_2/FeS, MoS_2/FeS/Rhe could effectively eradicate *Staphylococcus aureus*-induced osteomyelitis in the tibias of rats under medical MW irradiation. This MW-reactive nanocomposite, which integrated inorganic materials and herbal components, could potentially address the challenge of efficiently treating infections in deep tissues.

The notable efficacy of microwave hyperthermia primarily stems from the superior absorption capacity of pathological tissues compared to normal tissues, enabling rapid heating and selective treatment. Owing to their distinctive properties, MoS_2 nanosheets hold substantial promise for microwave thermal therapy and can serve as ideal microwave-susceptible agents.

4. MoS_2 for drug delivery system

MoS_2, composed majorly of molybdenite, is capable of being cast as nanosheets with crystalline arrangements based on the S-Mo-S monolayers (usually from one to three). The presence of S atoms on the surface layer enhances MoS_2's ability to adsorb small molecules. These characteristics provide MoS_2 with significant benefits as a drug delivery system [13]. Moreover, 2D MoS_2 nanosheets can also be explored as drug delivery vehicles because of their planar structure and high surface-area-to-volume ratio [14]. MoS_2 possesses unique physical and chemical properties that make it an effective adsorbent for various biomolecules and drug molecules, enhancing its tumor-targeting capabilities and colloidal stability. Its high specific surface area allows for significant drug loading, which is crucial for chemotherapy.

Zhihuai Chen et al. designed a delivery system based on MoS_2 that is modified with PEG and biotin and concurrently carried Cur and Er for enhanced cancer treatment (**Figure 4**) [15]. The system efficiently transformed NIR radiation into thermal energy, leading to the direct destruction of cancer cells through photothermal ablation and the regulated discharge of the chemotherapeutic agents Cur and Er. The biotin component enabled precise tumor targeting, ensuring the preferential concentration of the MoS_2-PEG-Biotin-Cur/Er complex at the cancerous region, which amplified the collaborative anticancer impact of Cur and Er. Notably, under NIR light exposure, MoS_2-PEG-Biotin-Cur/Er synergized chemotherapeutic actions with PTT, significantly inhibiting the spread of lung cancer cells and impeding tumor growth *in vivo*. Additionally, the research demonstrated that the synthesized MoS_2-PEG-Biotin-Cur/Er composite possessed superior tumor-fighting capabilities. Yun Liu et al. synthesized MoS_2/graphene oxide (GO) nanocomposites that address key challenges in nanomedicine [16]. These nanocomposites demonstrated superior dispersion in watery mediums and exhibited satisfactory biocompatibility in both *in vitro* and *in vivo* compared to GO alone. They selectively targeted the lungs and had an enhanced drug-loading capacity. The MoS_2/GO composites also showed increased tumor-killing efficacy against cancer cells, particularly those prone to metastasize to the lungs, and significantly suppressed metastatic tumor growth in mice. The study revealed that MoS_2/GO reduced reactivity with macrophages compared to GO, which affected the

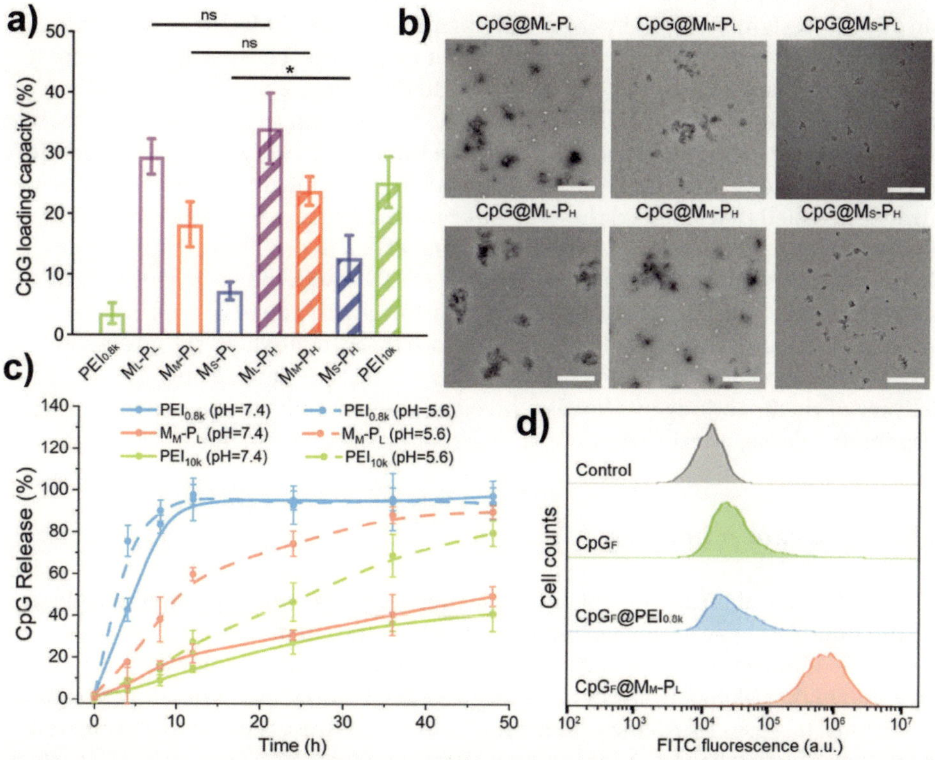

Figure 4.
MoS_2-based drug delivery system for delivering CpG adjuvant. Permission adapted from [16].

interaction and uptake of nanosheets. This discovery highlighted the potential of MoS_2/GO nanocomposites in cancer nanotheranostics, offering a multifunctional approach to treatment. In the aforementioned two papers, the synthesized nanocomposite materials have both improved targeting and stronger anti-tumor capabilities.

Hongxia Li et al. developed a novel nanoplatform using polyethyleneimine-coated MoS_2 nanosheets varying in dimensions and electrical charge concentrations to wrap CpG, a toll-like receptor-9 agonist, for the treatment of head and neck squamous cell carcinoma (HNSCC) [17]. These nanosheets, particularly those of medium size and low charge density, demonstrated efficient CpG loading and low cytotoxicity due to the flexibility and crimpability of 2D materials. The CpG-loaded nanosheets (CpG@MM-PL) were shown to enhance the maturation and function of dendritic cells and cytotoxic T lymphocytes *in vivo*, thereby improving the tumor immune microenvironment. The combination of CpG@MM-PL with ICB therapy, specifically anti-PD-1, significantly enhanced the therapeutic effect against HNSCC. This study not only provided a more effective strategy for HNSCC treatment but also highlighted the importance of considering the unique properties of 2D materials in the design of future nanosheet-based therapeutic platforms for drug delivery and cancer immunotherapy. Zhongyang Kou et al. created cationic MoS_2-PEG-PEI by utilizing branched polyethyleneimine (PEI) and polyethylene glycol that had been modified with LA-PEG [18]. The amine groups on the positively charged nanomaterials could combine with the negatively charged small interfering RNA (siRNA). The findings indicated that the innovative nanocarrier displayed favorable biocompatibility, low cytotoxicity, and a high capacity for gene transport without being affected by serum, making it a promising candidate for gene delivery and therapeutic applications. Bang Lin Li et al. designed a layer-by-layer self-assembled stacked MoS_2 superstructure, resembling a testudo, for the delivery of cancer drugs [19]. These structures were self-assembled using DNA oligonucleotides and disassemble in response to ATP, a metabolic marker of cancer cells. The assembly into a complex, higher-order DNA/MoS_2 nanosheets structure was facilitated by a linker aptamer, which promoted the interlayer assembly process. The MoS_2 nanosheets, functionalized with DNA, formed a protective shell that shielded the drug but could be triggered to release it by ATP binding to the aptamers, leading to targeted and efficient drug release for chemotherapy. This system combined the specificity of DNA with the properties of MoS_2, offering a stimuli-responsive drug delivery platform with potential applications in nanomedicine and nanotechnology. These two articles highlighted the interaction between nucleic acids and nanomaterials. This interaction was not only used for the delivery of therapeutic agents but also involved the assembly of nanostructures and responsive release, offering new possibilities for precision medicine.

5. MoS_2 for cancer imaging

Early biological measurement and high-precision positioning of tumors will greatly improve the accuracy of tumor treatment. Thus, many imaging strategies have been designed based on nanomaterials to precisely detect tumor [20]. Studies have shown that MoS_2 has low cytotoxicity and good biocompatibility under certain conditions. It causes less damage to normal cells and tissues, reducing side effects on the patient's body when used for cancer imaging and improving safety. Besides, since MoS_2 can specifically interact with tumor cells or specific biomarkers in the tumor microenvironment or achieve specific recognition and binding to tumor cells through targeted modification, it can provide high-sensitivity and high-specificity signals during the imaging process, which is helpful for accurate detection and localization of tumors. MoS_2 and MoS_2-based nanocomposites are successfully introduced in PA imaging, near-infrared fluorescence imaging, and MR imaging.

A nanoprobe, designated as MoS_2@PDA–PEG/peptide (MPPF), capable of activating and visualizing furin activity from within the body, was engineered to offer both NIRF and radiometric PA imaging capabilities [21]. The MPPF nanoprobes were synthesized by conjugating Cy7-tagged furin substrate peptides to polydopamine (PDA)-coated MoS_2 nanosheets through covalent bonding. When furin cleaves the peptides, Cy7 molecules detach from the MPPF nanoprobes and regain their fluorescent properties, enabling the detection of furin activity with a significantly reduced limit of detection (LOD). Concurrently, the disruption of the fluorescence resonance energy transfer from Cy7 to MoS_2 nanosheets and the swift elimination of the diminutive Cy7 molecules result in distinct reduction rates for PA_{768} and PA_{900}. This disparity renders the ratio PA_{768}/PA_{900} a reliable marker for gauging the level of furin activity. To enhance the accessibility of two-photon fluorescence (TPF) imaging techniques for the bioimaging field, Carrie Sweet et al. introduced the creation of a water-soluble, antibody-linked MoS_2 quantum dot (QD) probe (**Figure 5**) [22]. This probe is designed for the specific targeting and imaging of LnCaP prostate cancer cells in the biological II window. The research indicated that MoS_2 QDs functionalized with anti-PSMA antibodies are capable of performing targeted two-photon imaging on viable prostate cancer cells when illuminated with 1064 nm light, thanks to their elevated two-photon luminescence, outstanding photostability, and superior biocompatibility. MoS_2 QDs have not only demonstrated an exceptionally high two-photon absorption

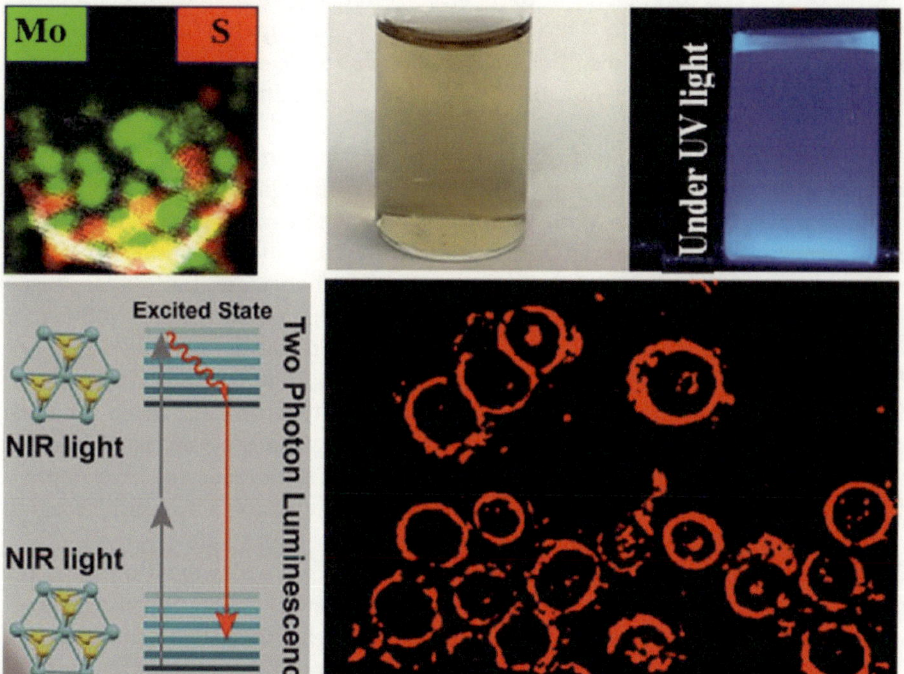

Figure 5.
MoS_2 quantum dot for cancer two-photon fluorescence (TPF) imaging. Permission adapted from [23].

cross-section and brightness but also the capability to differentiate between targeted and nontargeted cells through bioconjugation. This research elucidated the mechanism behind the superior two-photon brightness of MoS_2 QDs and offered a selective TPL agent based on zero-dimensional transition metal dichalcogenides for efficient imaging of live cells. In both articles, proteins were used to conjugate with MoS_2 to form nanomaterials, enhancing the corresponding imaging functions.

In addition, MoS_2 could be coupled with other molecules to form nanocomposite materials to improve the sensitivity and accuracy of imaging techniques. A novel 2D nanocomposite was developed by self-assembling functionalized iron oxide nanoparticles (IONPs) on the surface of two-dimensional MoS_2 nanosheets, forming MoS_2-IO nanocomposites [23]. These nanocomposites were further modified with two types of PEG to enhance stability in physiological environments. The commonly utilized positron-emitting isotope ^{64}Cu can be securely attached to the MoS_2 surface without requiring any chelating agents, thereby facilitating *in vivo* PET imaging. The robust NIR absorption and superparamagnetic properties of the MoS_2-IO-PEG nanocomposite enable its application in PA therapy and MR imaging. Leveraging the multimodal imaging techniques, which demonstrated the effective accumulation of MoS_2-IO-(d) PEG in tumors after intravenous injection, *in vivo* PTT was subsequently implemented, resulting in the successful ablation of tumors in an animal model. Chengbo Liu et al. introduced a MoS_2–Indocyanine Green (ICG) hybrid that was developed for the PA imaging of orthotopic brain glioma at deep sites [24]. This hybrid significantly enhanced PA imaging sensitivity compared to MoS_2 nanosheets alone. This linkage led to significantly elevated optical absorption across an extensive near-infrared range, a shift toward longer wavelengths in the absorption peak of ICG, and an improvement in the photothermal and PA conversion efficiency of ICG. Additionally, using MoS_2-ICG as a contrast agent for *in vivo* PA imaging of orthotopic glioma, the tumor mass could be visualized nearly twice as deep as compared to our previous imaging with molybdenum disulfide nanosheets. The MoS_2–ICG hybrid exhibited significant promise for efficient tumor molecular imaging in the field of translational medicine.

Some articles also mentioned that designing MoS_2 nanomaterials with specific structures or sizes could also bring significant effects. Jingqin Chen et al. developed a method to directly produce MoS_2 nanosheets featuring various layered nanostructures by utilizing the process of albumin-induced sequential exfoliation without further surface modifications [25]. Compared with few-layer and multilayer counterparts, single-layer MoS_2 exhibited higher imaging sensitivity, which was due to the enhanced NIR absorption and better elastic properties resulting in an amplified PA effect. The synthesized single-layer MoS_2 nanosheets exhibited minimal toxicity to cells, low hemolytic activity, robust cellular internalization, and enhanced PA imaging contrast. Utilizing single-layered MoS_2 nanosheets as potent contrast agents, *in vivo* PA imaging was successfully conducted in both subcutaneous and orthotopic brain tumor models, demonstrating exceptional sensitivity and substantial imaging depth. This represented the inaugural instance of employing single-layered MoS_2 with distinctive structural and functional attributes for PA imaging of brain tumors, highlighting its significant promise as a cutting-edge nanoplatform for advanced biomedical applications. Haifeng Dong et al. designed the controlled synthesis of small-sized MoS_2 QDs with expected optical properties [26]. These QDs were synthesized using a tetrabutylammonium-assisted ultrasonication method, which involved OH-mediated chain-like Mo-S bond cleavage of multilayered MoS_2 powder. The adjustable up-bottom method for the accurate synthesis of MoS_2 QDs ultimately facilitated thorough experimental studies on their optical characteristics. The fabricated MoS_2 QDs displayed excellent down-conversion photoluminescent properties and showcased significant up-conversion photoluminescence capabilities, suitable for bioimaging applications. Additionally, the MoS_2 QDs were proven to have a superior capacity for generating singlet oxygen compared to the commercial photosensitizer PpIX, indicating their promising potential in the field of photodynamic therapy.

6. Conclusions

In summary, MoS_2 has emerged as a highly promising material in the field of cancer diagnosis and treatment. Their unique X-M-X layered structure, along with the covalent bonding within layers and weak Van der Waals forces between layers, endows them with exceptional properties such as high photothermal conversion efficiency, substantial surface area, and biocompatibility, and offers new opportunities for multimodal treatment strategies. Despite the great advantages of MoS_2 in cancer therapy, several challenges remain. First, the synthesis process of MoS_2 needs to be further optimized to ensure its uniformity, reproducibility, and scalability for clinical applications. Second, the biosafety of MoS_2 requires in-depth study, especially in terms of long-term toxicity, biodistribution, and metabolic pathways. Additionally, the stability, catalytic efficiency, and photothermal conversion efficiency of MoS_2 in the tumor microenvironment also need to be further improved. Future research should focus on further optimizing the properties and performance of these materials, exploring their combination with other therapeutic modalities, and conducting more comprehensive preclinical and clinical studies to fully realize their potential in improving cancer treatment outcomes. Continued innovation and development in this area hold great promise for the future of cancer therapeutics.

Acknowledgements

This research was supported by the Beijing Municipal Natural Science Foundation (2254093), the Postdoctoral Fellowship Program of CPSF (GZC20230196), the project of Science and Technology of Yinchuan (2024SFZD004), and the union project of BUCT-CJFH biomedical center (PT2407, 2024-NHLHCRF-YXHZ-MS-05). The authors would like to thank the Fundamental Research Funds for the Central Universities (buctrc202419) for the support.

Conflict of interest

The authors declare no conflict of interest.

Author details

Wanting Yang[1,2] and Wensheng Xie[1,2*]

1 The State Key Laboratory of Organic–Inorganic Composites, College of Life Science and Technology, Beijing University of Chemical Technology, Beijing, P. R. China

2 The Beijing Laboratory of Biomedical Materials, Beijing University of Chemical Technology, Beijing, P. R. China

*Address all correspondence to: xws@mail.buct.edu.cn

References

[1] Manzeli S, Ovchinnikov D, Pasquier D, Yazyev OV, Kis A. 2D transition metal dichalcogenides. Nature Reviews Materials. 2017;**2**:1-15. DOI: 10.1038/natrevmats.2017.33

[2] Nandy SK, Das S, Pandey S, Kalita P, Gupta MK, Kabra A, et al. The futuristic applications of transition metal dichalcogenides for cancer therapy. Luminescence. 2024;**39**:e4771. DOI: 10.1002/bio.4771

[3] Srivastava M, Banerjee S, Bairagi S, Singh P, Kumar B, Singh P, et al. Recent progress in molybdenum disulfide (MoS_2) based flexible nanogenerators: An inclusive review. Chemical Engineering Journal. 2024;**480**:147963. DOI: 10.1016/j.cej.2023.147963

[4] Yu X, Xu C, Sun J, Xu H, Huang H, Gan Z, et al. Recent developments in two-dimensional molybdenum disulfide-based multimodal cancer theranostics. Journal of Nanbiotechnology. 2024;**22**:515. DOI: 10.1186/s12951-024-02785-x

[5] Fu C, Tan L, Ren X, Wu Q, Shao H, Ren J, et al. Interlayer expansion of 2D MoS_2 nanosheets for highly improved photothermal therapy of tumors in vitro and in vivo. Chemical Communications. 2018;**54**:13989-13992. DOI: 10.1039/C8CC08279K

[6] Chen L, Zhou X, Nie W, Feng W, Zhang Q, Wang W, et al. Marriage of albumin–gadolinium complexes and MoS_2 nanoflakes as cancer theranostics for dual-modality magnetic resonance/photoacoustic imaging and photothermal therapy. ACS Applied Materials and Interfaces. 2017;**9**:17786-17798. DOI: 10.1021/acsami.7b04488

[7] Liu T, Wang C, Cui W, Gong H, Liang C, Shi X, et al. Combined photothermal and photodynamic therapy delivered by PEGylated MoS_2 nanosheets. Nanoscale. 2014;**6**:11219-11225. DOI: 10.1039/C4NR03753G

[8] Chen L, Feng Y, Zhou X, Zhang Q, Nie W, Wang W, et al. One-pot synthesis of MoS_2 nanoflakes with desirable degradability for photothermal cancer therapy. ACS Applied Materials and Interfaces. 2017;**9**:17347-17358. DOI: 10.1021/acsami.7b02657

[9] Xie W, Gao Q, Wang D, Guo Z, Gao F, Wang X, et al. Doxorubicin-loaded Fe3O4@MoS_2-PEG-2DG nanocubes as a theranostic platform for magnetic resonance imaging-guided chemo-photothermal therapy of breast cancer. Nano Research. 2018;**11**:2470-2487. DOI: 10.1007/s12274-017-1871-1

[10] Wang S, Tan L, Liang P, Liu T, Wang J, Fu C, et al. Layered MoS_2 nanoflowers for microwave thermal therapy. Journal of Materials Chemistry B. 2016;**4**:2133-2141. DOI: 10.1039/C6TB00296J

[11] Fu C, He F, Tan L, Ren X, Zhang W, Liu T, et al. MoS_2 nanosheets encapsulated in sodium alginate microcapsules as microwave embolization agents for large orthotopic transplantation tumor therapy. Nanoscale. 2017;**9**:14846-14853. DOI: 10.1039/C7NR04274D

[12] Jin L, Liu X, Zheng Y, Li Z, Zhang Y, Zhu S, et al. Interface polarization strengthened microwave catalysis of MoS_2/FeS/Rhein for the therapy of bacteria-infected osteomyelitis. Advanced Functional Materials. 2022;**32**:2204437. DOI: 10.1002/adfm.202204437

[13] Liu J, Lu K, Gao F, Zhao L, Li H, Jiang Y. Multifunctional MoS_2 composite nanomaterials for drug delivery and synergistic photothermal therapy in cancer treatment. Ceramics International. 2022;**48**:22378-22386. DOI: 10.1016/j.ceramint.2022.04.244

[14] Zhao Y, Wei C, Chen X, Liu J, Yu Q, Liu Y, et al. Drug delivery system based on near-infrared light-responsive molybdenum disulfide nanosheets controls the high-efficiency release of dexamethasone to inhibit inflammation and treat osteoarthritis. ACS Applied Materials and Interfaces. 2019;**11**:11587-11601. DOI: 10.1021/acsami.8b20372

[15] Chen Z, Wei X, Zheng Y, Zhang Z, Gu W, Liao W, et al. Targeted co-delivery of curcumin and erlotinib by MoS_2 nanosheets for the combination of synergetic chemotherapy and photothermal therapy of lung cancer. Journal of Nanbiotechnology. 2023;**21**:333. DOI: 10.1186/s12951-023-02099-4

[16] Liu Y, Peng J, Wang S, Xu M, Gao M, Xia T, et al. Molybdenum disulfide/graphene oxide nanocomposites show favorable lung targeting and enhanced drug loading/tumor-killing efficacy with improved biocompatibility. NPG Asia Materials. 2018;**10**:e458-e458. DOI: 10.1038/am.2017.225

[17] Li H, Liu M, Zhang S, Xie X, Zhu Y, Liu T, et al. Construction of CpG delivery nanoplatforms by functionalized MoS_2 nanosheets for boosting antitumor immunity in head and neck squamous cell carcinoma. Small. 2023;**19**:2300380. DOI: 10.1002/smll.202300380

[18] Kou Z, Wang X, Yuan R, Chen H, Zhi Q, Gao L, et al. A promising gene delivery system developed from PEGylated MoS_2 nanosheets for gene therapy. Nanoscale Research Letters. 2014;**9**:587. DOI: 10.1186/1556-276X-9-587

[19] Li BL, Setyawati MI, Chen L, Xie J, Ariga K, Lim C-T, et al. Directing assembly and disassembly of 2D MoS_2 nanosheets with DNA for drug delivery. ACS Applied Materials and Interfaces. 2017;**9**:15286-15296. DOI: 10.1021/acsami.7b02529

[20] Bai J-W, Qiu S-Q, Zhang G-J. Molecular and functional imaging in cancer-targeted therapy: Current applications and future directions. Signal Transduction and Targeted Therapy. 2023;**8**:1-32. DOI: 10.1038/s41392-023-01366-y

[21] Li X, Xiu W, Xiao H, Li Y, Yang K, Yuwen L, et al. Fluorescence and ratiometric photoacoustic imaging of endogenous furin activity *via* peptide functionalized MoS_2 nanosheets. Biomaterials Science. 2021;**9**:8313-8322. DOI: 10.1039/D1BM01410B

[22] Sweet C, Pramanik A, Jones S, Ray PC. Two-photon fluorescent molybdenum disulfide dots for targeted prostate cancer imaging in the biological II window. ACS Omega. 2017;**2**:1826-1835. DOI: 10.1021/acsomega.7b00229

[23] Liu T, Shi S, Liang C, Shen S, Cheng L, Wang C, et al. Iron oxide decorated MoS_2 nanosheets with double PEGylation for chelator-free radiolabeling and multimodal imaging guided photothermal therapy. ACS Nano. 2015;**9**:950-960. DOI: 10.1021/nn506757x

[24] Liu C, Chen J, Zhu Y, Gong X, Zheng R, Chen N, et al. Highly sensitive MoS_2–indocyanine green hybrid for photoacoustic imaging of orthotopic brain glioma at deep site. Nano-Micro Letters. 2018;**10**:48. DOI: 10.1007/s40820-018-0202-8

[25] Chen J, Liu C, Hu D, Wang F, Wu H, Gong X, et al. Single-layer MoS_2 nanosheets with amplified photoacoustic effect for highly sensitive photoacoustic imaging of orthotopic brain tumors. Advanced Functional Materials. 2016;**26**:8715-8725. DOI: 10.1002/adfm.201603758

[26] Dong H, Tang S, Hao Y, Yu H, Dai W, Zhao G, et al. Fluorescent MoS_2 quantum dots: Ultrasonic preparation, up-conversion and down-conversion bioimaging, and photodynamic therapy. ACS Applied Materials and Interfaces. 2016;**8**:3107-3114. DOI: 10.1021/acsami.5b10459